William Martin

The Illustrated Natural Philosophy

Being a Manual of Modern Science, for Schools and Families

William Martin

The Illustrated Natural Philosophy
Being a Manual of Modern Science, for Schools and Families

ISBN/EAN: 9783337025267

Printed in Europe, USA, Canada, Australia, Japan

Cover: Foto ©Thomas Meinert / pixelio.de

More available books at **www.hansebooks.com**

THE ILLUSTRATED
NATURAL PHILOSOPHY;
BEING A
MANUAL OF MODERN SCIENCE,
FOR SCHOOLS AND FAMILIES.

WITH NUMEROUS ILLUSTRATIONS, QUESTIONS,
AND EXPERIMENTS.

By WILLIAM MARTIN,
AUTHOR OF THE "PHILOSOPHY OF TEACHING," "EDUCATIONAL MAGAZINE,"
"HOLIDAY BOOK," ETC.

EIGHTH EDITION.

LONDON:
SIMPKIN, MARSHALL, & Co., STATIONERS' HALL COURT;
HAMILTON, ADAMS, & Co., PATERNOSTER ROW.

1864.

LONDON:
J. AND W. RIDER, PRINTERS,
14, BARTHOLOMEW CLOSE.

PREFACE.

The importance of NATURAL PHILOSOPHY as a study is universally admitted; and a popular MANUAL OF MODERN SCIENCE, comprehending the great discoveries of the day, divested of absurd theories and useless technicalities, carefully arranged and freely illustrated, has long been a desideratum both to parents and teachers; and to none more than to that increasing number of EDUCATORS who are anxious to teach not only *words* but *things;* and who, in addition to storing the mind with the facts of knowledge, would elaborate that knowledge by such a process of intellectual training, as will teach their pupils to *observe, reflect, compare,* and *judge.*

To develope, then, this great principle of Education, which the author ventures to call "THE INTELLECTUAL METHOD," this Manual has been prepared, upon a plan essentially different from that of any other book. It is divided into *chapters* and *sections*, that it may be read *in class,* or used for *individual study;* while at the same time it possesses all the advantages of the *catechetical* form, without the defect of the *set* question and *set* answer system, equally derogatory to those who *learn* and those who *teach* it. ILLUSTRATIONS are far more numerous and striking than in any existing work on the subject, and are not confined merely to the text, but explain an enlarged series of additional facts, of a highly interesting character, which, while they are exponents of the general principles of the sciences treated on, are also examples of their successful application.

With regard to the *interrogative portion* of the work, the reader will perceive that the chapters and sections are all *numbered,* so as to correspond with an enlarged series of questions, which may be answered either *orally* or in *writing*. The paragraphs of the text are *written* with a view to questioning, and each sentence has been tested and proved in the school room; while the questions themselves are prepared in such a form as to induct the teacher in the art of questioning. Hence they progressively ascend from particulars to generals; those at the latter part of the course being little more than skeleton exercises, capable of amplification at the discretion of the examiner, and leading to that most important part of all instruction, "*Questioning upon the answers* of the children."

Ten years experience as a teacher, and subsequently twelve as a public writer on education, give the author some claim to be heard on this subject; and he would beg leave to impress upon the tutor the importance of rigid interrogation on every branch of knowledge, with a view to the exercise of *all the faculties*. This principle fully carried out, although it may appear difficult at first, will in the end reward him for all his labours, by the ease with which subsequent knowledge will be acquired, and by the high satisfaction he will feel in reflecting that he has *done his duty to his pupils*.

The following is a specimen of the method of questioning, which might be adopted as introductory to this work, and may serve as an example of the manner in which questions should be put.

ARTICLES FOR ILLUSTRATION.

A WHITE FLINT—PIECE OF CLAY—INDIA RUBBER BALL—WATCH-GLASS—WATER AND TUMBLER—LAMP OR TAPER—STEEL—BIRD'S WING—BELL, &c.

Teacher.—What is this? *Flint.*—What is its colour? *White.*—Its form? It may be *round, roundish, oval, oblong.*—Take it and feel it. Now tell me something more about it? It is *hard*, it is *rough*, some portions *project*, others are *indented*, slightly *concave* or *hollow*.—Explain concave and convex from watch-glass. What would you call this proportion of the flint? A projection, a protuberance.—Explain the words. Can you see through the flint; why not?—*The light will not pass through it.*—Will it pass through this glass. What is a body called which the light passes through? *Transparent.*—What is a body called which will not suffer the light to pass through it? *Opaque.*—Mention substances transparent and opaque; explain *semi-transparent* from a sheet of writing paper. Have you seen the whole of the flint? No.—What part have you seen? The outside.—What part have you not seen? The inside.—Another word for inside; what shall I do that you may see the inside? Break the flint.—I have broken the flint; is it the same as it was? No; it has many parts.—What does this experiment prove? That the flint has divisibility; it can be divided into parts.—Are these parts of the same form and colour as the flint? The parts that were inside are *black*.—Are they unlike the outside in any other particular? They are smooth, the outside was rough.—In what other particulars are they unlike? They have sharp edges and points, and shining sides.—What form are they? They are angular, jagged, &c.—Now tell me by which of your senses you discovered the flint to be *white;* by which sense did you discover it to be cold and hard; by which of your senses did you discover it was sharp? By the eye?—Could you discover this by any other sense? By the touch.—Now shut your eyes, and tell me if you can discover it to be hard without touching it? (Teacher strikes two pieces of the broken flint together).—What kind of sound do you hear? A sharp clashing sound.—Do soft bodies make such a sound? Do all hard bodies make a sharp clashing sound when struck? Some make a dull or heavy sound. (Strikes two pieces of lead together).—What kind of sound has a bell? It is said to be sonorous. (Strikes a bell).—How is the sound produced? By *vibration.*—Explain vibration.—Teacher should here exercise the ear by striking various bodies together, the pupils keeping their eyes closed.

Teacher pouring out water.—What is the difference between the flint and the water? One is solid, the other is *fluid.*—If I put the flint into the water, what happens? If I put a cork into the water, what happens? Why does the stone sink and the cork float? Because one is heavier than the water, and the other is lighter.—How to you mean heavier than water? Than such a space of water as the flint and cork occupy?—What is the difference between the *flint* and the *cork?* To which of the kingdoms of nature does each belong? What is the difference between a mineral, a vegetable, and an animal? An animal *lives, grows, feels,* and MOVES; a vegetable *lives* and *grows;* a mineral neither lives nor grows, it *exists.*—How would you express the difference between a tree and a flint? The tree is *organised,* the flint is *inorganized.*—The principal organs of a plant are the *roots, leaves, flowers,* &c.; those of the animal, the heart, brain, eye, limbs, &c.

You told me that flint was black; here is a small thin piece, is this black? No; grey.—Is it opaque? No.—Is it transparent; can you see through it? Yes; the light.—It is partly transparent; this is called semi-transparent.—You told me that the flint was white on the outside; I will scrape it. What is it now? Black.—Was the white substance a part of the flint?—*(They will hesitate at this question).*— No, it was not. You erred when you said the outside of the flint was *white;* it was a substance on the outside. You erred, also, when you said the inside was *black*. What should this teach you? Not to trust too much to our senses.—What is this white part; is it different to the flint? Yes.—In what particular? The flint is *hard;* the white substance is soft, and *crumbles*.—Yes; it is what is called *friable*. The outside of the flint is composed of *chalk*. Shut your eyes, and tell me, by touching it with the tongue, if there be any difference in their taste, between the flint and the chalk? Yes; the flint has no taste; the chalk is slightly bitter, and sticks to the tongue.—The flint, then, is *insipid* (tasteless); the chalk is *bitter* and *adhesive*. The chalk and flint are not *the same*.

Can I burn this flint? Yes.—How? *(By putting it in the fire, will probably be the answer).*—Can I burn it in any other way? Do you think I could make it burn itself, without applying any fire to it? Feel the flint; is it warm or cold? Cold.—I will rub it on the floor; feel it now? It is hot.—What produced the heat? The rubbing.—What is the rubbing called? *Friction.*—What is this? A piece of steel.—Now, what do you see? *(Strikes the flint and steel together).*—Sparks of fire.—What are they composed of? Small pieces of flint, heated red hot.—By what? *Friction*—No; by *percussion*.—The word *percussion* is used when a body is *struck; friction*, when it is *rubbed.*—What is the dust on this paper? The ashes of the flint.—What has happened to the flint? It has been burnt.

Such is an imperfect specimen of the manner in which the " INTELLECTUAL METHOD" of interrogation should be carried out. Its *principle* is the exercise of the *observing* and *reflecting* faculties, in agreement with the laws which regulate the human mind in the attainment of knowledge; its *object* is to enable the pupil to conceive of things *clearly* in all their *particulars, comprehensively* in all their *generals,* and *systematically* in all their *relations*. Thus the pupil is, in the strictest sense of the word, *educated;* the EDUCATOR becomes *elevated,* and EDUCATION itself is raised TO THE RANK OF A SCIENCE.

WILLIAM MARTIN.

Preface

TO THE

THIRD EDITION.

In presenting the THIRD EDITION of the Illustrated Natural Philosophy to the Public, the Author cannot refrain from expressing his satisfaction at the rapid sale of the last, as a proof that an Intellectual method of teaching is taking the place of that absurd *parrot work*, which is especially exemplified in Catechisms, and so called Interrogative books, which teach words and not things, empty sounds and unmeaning phrases, in place of correct ideas and real knowledge: and by which subjects are said to be *learned*, but are not remembered; *committed* to memory, but not known.

In the present edition, the Author has spared neither labour nor expense, to make the work in every way worthy of the distinguished patronage it has received. It has been thoroughly revised, corrected, and improved, and very considerably increased, by the addition of a numerous series of practical experiments in all the sciences; and now presents a body of information, upon elementary science, far more complete and succinct than is to be found in any other work of its size and character; especially adapted to exercise and strengthen the thinking powers, to excite the pupil to investigation and research, and inspire him with a thirst after Knowledge and a love of Truth.

CONTENTS.

LAWS OF MATTER AND MOTION.

Chapter		Page
I.	Introduction	1
II.	Cause and Effect	4
III.	Properties of Matter	5
IV.	Subordinate Properties	10
V.	Motion, and its Laws—Time	13
VI.	Attraction	18
VII.	Centre of Gravity, and Central Forces	20

MECHANICS.

VIII.	Mechanical Powers, (Levers)	25
IX.	Wheel, Axle and Pulley	30
X.	Inclined Plane, Wedge, and Screw	37
XI.	Resisting Mediums—Fly Wheel and Friction	42
XII.	The Pendulum	47

HYDROSTATICS AND HYDRAULICS.

XIII.	Pressure of Fluids	51
XIV.	Specific Gravity	57
XV.	Laws of Hydraulics	62
XVI.	Pumps	63
XVII.	Hydraulic Machinery	69

PNEUMATICS.

XVIII.	Physical Properties of the Air	73
XIX.	Air Pump and Experiments	84
XX.	Barometer, Thermometer, and Hygrometer	94
XXI.	Æronautics and Diving Bell	100

METEOROLOGY.

XXII.	Meteorological Phenomena — Evaporation, Rain, Hail—Dew—Clouds—Mists—Winds — Monsoons Hurricanes—Tornadoes—Waterspouts	105

ACOUSTICS.

XXIII.	Propagation of Sound—the Animal Ear, &c	110
XXIV.	Reflection of Sound	115
XXV.	Melody and Harmony—Musical Instruments.	117

CONTENTS.

OPTICS.

Chapter		Page
XXVI.	Theories and Phenomena of Light	126
XXVII.	Dioptrics	128
XXVIII.	Catoptrics	132
XXIX.	Refraction by Prisms and Lenses	137
XXX.	Refraction of Light by Prisms and Lenses	138
XXXI.	The Prism—Chemistry of Light, &c.	140
XXXII.	The Rainbow	144
XXXIII.	The Eye	146

ASTRONOMY.

XXXIV.	History of Astronomy	156
XXXV.	Solar System	159
XXXVI.	Mercury and Venus	163
XXXVII.	Earth and Moon	165
XXXVIII.	The Moon's Motions	167
XXXIX.	Mars and the Asteroids—Jupiter—Saturn — Herschel and Neptune	170
XL.	The Tides and Eclipses	176
XLI.	Fixed Stars—Nebulæ—Comets, &c.	179

CHEMISTRY.

XLII.	Nature of Chemistry	183
XLIII.	Of the Elements of Bodies	186
XLIV.	Oxygen—Hydrogen, and Nitrogen	188
XLV.	Chlorine—Iodine—Bromine—Fluorine	192
XLVI.	Carbon—Sulphur—Phosphorus	193
XLVII.	Earths	196
XLVIII.	Metals	199
XLIX.	Acids and Salts—Toxicology. &c.	204
L.	Heat, or Caloric	208

ELECTRICITY.

LI.	Derivation—Electric Action—Electrical Machine—Theories of Electricity — Excitation — Distribution—Attraction and Repulsion—Induction—Leyden jar — Electrometer—Universal Discharger— De Luc's Column—Animal Electricity	212

GALVANISM.

LII.	Discovery by Sulzer—Galvani—Mode of Excitation—Galvanic Pile — Trough Battery — Chemical, Electrical, and Physiological Effects	225

MAGNETISM.

LIII.	Theory—Natural Magnet—Artificial Magnet—Mariner's Compass — Polarity — Azimuth Compass—Dip of the Needle—Variation — Animal Magnetism—Electro-Magnetism—Thermo-Electricity	231

ILLUSTRATIONS	237
QUESTIONS	277
EXPERIMENTS	297

Natural Phenomena.

NATURAL PHILOSOPHY.

CHAPTER I.

DEFINITION.—THE UNIVERSE.—SPACE, ABSOLUTE AND RELATIVE.—MATTER.—ATTRACTION.—REPULSION.—CAUSE AND EFFECT.

1. NATURAL PHILOSOPHY, which is also called PHYSICS, from the Greek word Φυσις, Phusis, signifying NATURE, embraces the whole UNIVERSE. It teaches the general pro-

perties of MATTER, and the action of *material substances* upon each other. It enables us to investigate and explain NATURAL PHENOMENA, to trace EFFECTS from CAUSES, and to ascertain those FIXED LAWS by which all things are sustained and regulated, called the LAWS of NATURE.

THE UNIVERSE.

2. The UNIVERSE consists of SPACE, i. e., unlimited *extension*, and of MATTER in *space*. The mind cannot limit the dimensions of the UNIVERSE. Thousands of stars which we behold on a dark night, are but a speck in it, and even some millions of suns and planets are discovered by the aid of the telescope, beyond the limits of natural vision.

SPACE.

3. SPACE is an abstract idea. It is either *absolute* or *relative*.

4. ABSOLUTE SPACE is mere extension; it has no limits or bounds, and can be described only by its want of properties.

5. RELATIVE SPACE is that part of absolute space which is occupied by any body, and is compared with any other part occupied by another body.

MATTER.

6. MATTER or *Substance* is every being or thing that acts upon the senses, either immediately, or by the perceptible effects it produces.

7. MATTER is composed of a concentration of small particles, called ATOMS, united in regular crystalized forms, so as to compose a compact mass. Six primitive regular forms can be traced; but these form infinite varieties.—*(See Frontispiece.)* Read Illustration No. 1.

8. These primitives are, the *parallelopipedon*, which includes the *cube*, the *rhomb*, and all the solids terminated by six faces paralleled two and two; the *tetrahedon*; the *octahedron*; the *hexahedral*, or *six-sided prism*; the *dodecahedron*, with equal and similar *rhomboidal planes;* and the *dodecahedron* with *triangular planes*.

9. Some kinds of matter are *visible*, such as wood, stone, &c., which depends upon the property they possess of not permitting the light to pass through them. Other bodies are

always *invisible*, and their existence is to be ascertained only from their effects upon other substances: of this nature is the atmospheric air by which we are surrounded, which, although perfectly invisible, is nevertheless as much a substance or matter as the hardest stone.

10. The power which causes the atoms or particles of matter to unite and form a body more or less solid, is called ATTRACTION. This power pervades the whole of nature, though certain bodies, under particular circumstances, possess an opposite power called REPULSION, that causes them to separate from each other. It is to be remembered too, that all bodies separate from each other by the repulsion of *heat* or *caloric*.

11. Thus the universe, and all that it contains, that ever has been, or ever may be made apparent to our minds, through the medium of the organs of sense, is composed of very minute indestructible atoms, called *matter*, which, by mutual attraction, cohere, or cling together, in masses of various form and magnitude.

12. These atoms are more or less near to each other, according to the *repulsion of heat* among them; and hence they exist under three general forms, or states,—*solid, liquid*, and *aëriform*, as in crystal, water, and air. In the first of these states the atoms are mutually fixed and combined. In the second, the atoms move freely upon each other and among themselves; and in the third, they are removed to a greater distance from each other. (See Chemistry and Pneumatics.)

13. Certain modifications of attraction and repulsion produce the subordinate peculiarities of state, called *porous, dense, hard, elastic, brittle, malleable, ductile*, and *tenacious*.

14. The states and substances to which we have referred are, generally, *inorganic*. But matter is also *organic;* that is, disposed into *organs*, which perform certain *functions* in the *animal* and *vegetable* economy necessary to *life*. Man is an *organized* being,—the eye is his organ of sight. A tree is an *organized* substance,—the leaf is the organ of respiration in the tree. The consideration of organized matter is called animal and vegetable *Physics*.

CHAPTER II.

OF CAUSE AND EFFECT.

CAUSES, PRIMARY AND SECONDARY.—KNOWLEDGE OF CAUSES —KNOWLEDGE OF EFFECTS.—THE SCIENCE OF CAUSATION —ADVANTAGE OF A KNOWLEDGE OF FACTS.

15. EVERY thing that exists must have a cause. A cause is either *primary* or *secondary*. The primary, or GREAT FIRST CAUSE of all things, is GOD, the *Creator*, the *Preserver*, and *Ruler* of all; and all that is can only be referred to eternal causation.

16. *Secondary*, or *proximate* causes, are those which immediately relate to an effect, as the sun is the cause of light and heat. The rotation of the earth on its axis is the cause of day and night.

17. It is of great importance to us that we should understand the cause of every thing brought under our observation. By knowing the causes of certain effects, we are often enabled to avert the danger to be apprehended from them.

18. A knowledge of *effects* is also of equal consequence to us, as effects become causes. The marksman calculates the effect of so much powder, such a weight of ball, such an elevation of his gun, the resistance of the air, the force of gravity: he fires, and hits the *mark*.

19. Natural Philosophy is peculiarly the science of causation; it inquires not only *what*, but *how* is this; and its perfection consists in the detection of the *nearest*, or *proximate* cause of things. It calculates effects both near and remote; and thus the human mind, *foreseeing* and *counteracting*, obtains in degree the mastery over matter, and approaches, though at humble distance, the wisdom of its Creator.

20. To ascertain the nature of causes, and their effects, the philosopher proceeds upon experience, or the *induction of*

FACTS. The larger the number of *facts* present to the mind in the investigation of any *truth*, the greater the probability of success; the more we know of *facts*, the more we shall know of *causes* and *effects*; of *Nature*, and of GOD.

CHAPTER III.

OF THE GENERAL PROPERTIES OF MATTER.

The *Essential* properties of matter are two:—EXTENSION, and SOLIDITY or IMPENETRABILITY.

Divisibility, Mobility, Inertia, Hardness, Elasticity, Brittleness, Malleability, Ductility, Tenacity, and *Gravity,* are *Non-essential* properties.

21. EXTENSION is that property of matter by which it occupies space.

SOLIDITY is that property of matter by which two bodies cannot occupy the same place at the same time.

22. If in a full glass of water a stone be placed, the water will be forced over. Some kinds of matter appear to be, but are not, exceptions to this rule. Salt may be added to water in considerable quantities, without making it run over. This arises from the particles or atoms of the water being of a globular form, and the particles of salt finding their way between them.

Globules.

23. Some substances, when mixed together, take up less room than they did in separate states. A glass of water, and

---

* By *solidity* or *impenetrability* in common language is understood the property of not being easily separated into parts, and, therefore, we must be careful not to confound this meaning of the term with the property just mentioned.

a glass of spirits of wine, mixed together, will not afterwards be found twice to fill the glass.

24. Zinc and copper melted together, take up less room. This arises from the closer union of the particles, and by their interstices being filled up.

25. Most bodies when heated occupy more room than before. Hot water and hot air take up more room than when cold, because their particles move further from each other by the repulsion of heat. When water is converted into steam, it acquires nearly two thousand times its former bulk, as the atoms of which it is composed recede from each other; at the same time air may be compressed a hundred parts into one.

26. Thus all bodies are more or less dense or compact, and more or less porous. The more porous are generally light, such as sponge, wood, and cork; the less porous are heavy, such as gold, lead, or stone.

27. Solidity is, therefore, a property of matter, by which a certain quantity occupies a certain space. Water is the common standard to which dense bodies are referred, and hence we have what are termed *specific gravities.*—(*See Hydrostatics.*)

28. DIVISIBILITY is that property by which matter is capable of being separated into parts, which may be removed from each other.

29. The atoms which compose the various masses of matter, with which we are acquainted, are so minute as to escape our observation.

30. Of the extreme minuteness of the atoms which constitute matter we may judge by the following illustrations; which, although they by no means exhibit the atoms themselves, assist the mind in its efforts to conceive how minute the atoms of matter must be.

31. Leuwenhoeck proved by the microscope, that a grain of sand would cover one hundred and twenty-five orifices, through which we daily perspire; and gave an account of certain animalculæ, so small, that three hundred thousand of them were not equal to a grain of sand; and that a pound of them would comprise more living creatures than there are human beings on the face of this globe. Subjoined are the forms of

DIVISIBILITY. 7

some of them; and it must be remarked, that the organization of many of them is exquisitely perfect.

32. There is a little fungus, called the puff-ball, which emits a fine dust, so light, that it floats in the air like smoke. Every particle of this dust, which is the seed of the plant, contains an organized germ; and yet the diameter of a hair would be one hundred and twenty-five thousand times as great as the seed of the lycoperdon, or puff-ball.

Puff-ball.

33. Musk, in the quantity of a single grain, will perfume for years a chamber twelve feet square. Such a chamber contains more than two millions of cubic inches of air, and each cubic inch must contain many thousands of the particles of musk, and this air must have been changed many thousands of times.

34. Gold-beaters, by hammering, reduce gold to leaves the 360 thousandth part of an inch in thickness. Thus a grain

of gold hammered out will cover fifty square inches. It is computed that sixteen ounces of gold might be laid over enough of silver wire to circumscribe the whole earth.

35. A piece of silver, or any other metal, may be dissolved in an acid, or dissipated by intense heat into vapour; yet it can be recovered from these states, and collected again, to form the original mass of silver; and it would not be more wonderful for the particles of our own bodies to reunite after death, and become re-organized.

36. A pound of wool may be spun so fine as to extend ninety-five miles; the fibres of a pound of silk will extend five hundred and eighty-three miles; but the threads of the spider are still finer—several thousand of them go to form a single thread, and a pound of them would reach round the earth.

37. MOBILITY is that property of matter by which it is capable of being moved from one part of space to another.

38. It is found by experiment and observation that all matter is capable of being moved, if a sufficient force can be applied for that purpose. (*See Motion*).

39. INERTIA or *inactivity* is the tendency which bodies have to continue in the same state into which they are put, *whether of rest or motion*, unless prevented by some external force.

40. Let a small card be placed upon one of the fingers of the left hand, and over the card, and immediately above the finger, a small piece of money; if, then, a smart blow be given to the card by the fore or middle finger of the right hand, the card will pass from under the money, which will be left upon the finger.

41. If a thin smooth piece of wood be laid over two wine-glasses, and immediately above the glasses two pieces of money; then, upon giving one of the ends, a smart blow in

a lateral direction, the slip of wood will pass from under the pieces of money, which will accordingly fall into the glasses

42. We said that a body in motion had a tendency to continue in that motion; and, if we run a short distance, we shall find it difficult to stop on a sudden. Thus it is that unskilful horsemen, when a horse suddenly stops, are thrown over his head; or people riding in a coach are suddenly thrown forward, when the coach stops. In a hare-hunt, sometimes the hare, by the power of its instinct, doubles upon the dog, as it is called, and thus defeats him in the chase. At C, where the hare is hard pressed by the dog, she makes a sudden turn in a new direction; the dog, being unprepared for this, is carried forward by inertia, or the tendency which his body has to proceed in a straight course to D. At D he brings himself round, and pursues the hare in the new direction to E; at E, where the hare is again hard pressed, she makes another quick turn toward G, and the dog is, as before, thrown forward to F. Thus the dog loses ground upon the hare at every turn or double which she makes; and hence it is that she may at length reach the cover, B, and eventually escape.

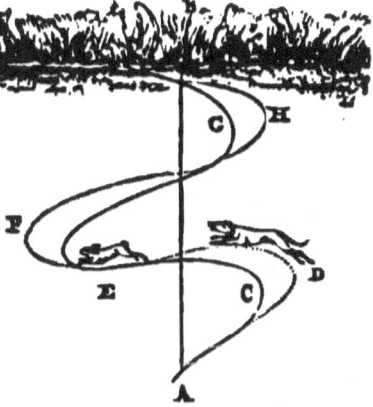

By this we may learn, that in all the avocations of life, much is to be learned from philosophy.

These are examples of the inertia of motion; 40 and 41 of the inertia of rest.

CHAPTER IV.

OF THE SUBORDINATE PROPERTIES OF MATTER.

43. Certain modifications of attraction and repulsion produce the subordinate peculiarities of state, such as Hardness, Elasticity, Brittleness, Malleability, Ductility, Tenacity, &c.

44. HARDNESS in bodies is not according to their densities. Flint is harder than lead, although lead is the more dense and heavy. Gold, though soft, is four times heavier than the diamond, which is the hardest body known. Hardness depends upon the force with which the atoms hold their places, or on some particular arrangement of them.

45. ELASTICITY in bodies is found when the atoms yield to a certain extent, when force is applied; but move back again, or regain their natural positions on the force being withdrawn.

46. The principle of elasticity is seen throughout the whole of the vegetable kingdom : it is great in the ash and willow, and slight in the various fungi. It is seen also in the animal and mineral departments.

Most hard bodies are elastic, as steel, glass, iron; and many soft ones, as india-rubber, silk, and a harp-string. All aeriform bodies are elastic.

47. A good steel sword may be bent until its ends meet, and, when allowed, will return to perfect straightness; a piece of cane may be twisted round, and will recover itself; and whalebone, from its elasticity, is applied in various articles of daily use,

Elasticity.

such as umbrellas, stays, brooms, &c.; while of steel the finest watch-springs are made.

48. Ivory, in its most compact state, is elastic. If an ivory ball be let fall on a marble slab, it will rebound nearly to the height from which it fell, and no mark be left on either. If the slab be wet, it shows a circular surface of some extent, dried by the blow; proving that extended parts of the ball must have come in contact with it, and recovered themselves afterwards by their elasticity.

49. BRITTLENESS belongs to most hard bodies, and particularly to those on which the cohesion of the atoms that compose them exists within narrow limits. The most brittle bodies are glass, china ware, cast-iron, and very hard steel.

50. MALLEABILITY in bodies arises from the willingness of their atoms to cohere in every direction. The term malleable means, that the body is capable of being reduced to thin plates by the hammer. Gold is extremely malleable, and it may be reduced to leaves so thin, that one thousand eight hundred would only be as thick as the leaf on which this is printed. Silver, copper, and tin, are also malleable to a considerable extent.

Gold-beater.

51. DUCTILITY means the susceptibility of bodies to be

Wire Drawing

drawn into wire, and it arises from the strong cohesion of their particles. *Platina*, *silver*, *copper*, and *gold*, are very ductile.

Dr. Wollaston produced platinum wire finer than the thread of the spider, to which I have before alluded.

52. Melted glass can be spun into threads so fine, as to be scarcely perceptible. This glass thread also proves the elasticity of the substance, for it may be made to resemble the beautiful hair and feathers of a bird, waving in the air like the plumes of a bird of Paradise.

53. TENACITY, or toughness, is seen in a great variety of bodies: in the metals, in the woods, and in many animal substances. The tendons of the animal body are exceedingly tough; the hair and wool of animals, twisted into threads, the silkworm's and spider's thread, hempen cord, &c. This quality in bodies has been of great importance in the arts of life.

54. The following table shows the comparative tenacity, or strength, to resist pulling, of certain metals and woods upon rods, whose diameter would be one-thousandth part of a square inch :—

	lbs.		lbs.
Cast-steel	134	Tin	5
Wrought-iron	70	Lead	2
Cast-iron	19	Teak	13
Copper	19	Oak	12
Platina	16	Birch	12½
Silver	11	Ash	14
Gold	9	Deal	11

Steel wire will support about thirty-nine thousand feet, that is, seven and a half miles, of its own length. One of the most splendid illustrations of the tenacity of iron is to be found in the Menai bridge.

CHAPTER V.

OF MOTION AND ITS LAWS.

55. MOTION is a simple idea, and, therefore, admits not of definition. A Grecian philosopher being asked to define *motion*, began to walk, and replied, this is motion. Motion, therefore, signifies a *change of place*, and implies the necessity of *time* as well as *space*. Its original primal cause is in God.

56. There are *two kinds of motion*, 1st. That by which an entire body is conveyed from one place to another, as when a stone falls from the hand to the ground, or a ship sails from port to port: 2nd, The imperceptible motion of the parts of bodies among themselves, which relate to the various compositions and decompositions of matter. (See Chemistry).

57. In the language of mechanical philosophy, the cause of any *change of motion* is called a moving or a changing FORCE. The words, therefore, *force* and *power*, denote that which causes *a change in the state of a body*, whether that state be rest or motion.

58. Motion may be *rapid*, as in the lightning; *slow*, as in the shadow of the sun-dial; *straight*, or *rectilineal*, as in the apparent path of a bullet dropped from the hand; *bent*, or *curvilinear*, as in the track of the same bullet fired from a gun; *uniform*, when the motion continues the same; *accelerated* when it increases; and *retarded*, when any thing tends to lessen its velocity.

59. The quantity, or force of motion in any moving body, is called its *momentum*. *Two bodies will move with the same force, if their velocities be to each other inversely, as their quanties of matter.*

60. A cannon-ball, of *a thousand ounces*, moving *one foot per second*, has the same quantity of motion in it as a musket-ball of *one ounce*, leaving the gun-barrel with the velocity of *a thousand feet* in the second.

61. From these general principles flow directly three axioms, which are usually called the *laws of motion*. They are, however, more properly the *laws of human judgment* with respect to motion.

LAWS OF MOTION.

62. FIRST LAW OF MOTION.—*Every body continues in a state of rest or of uniform rectilineal motion, unless affected by some extraneous force.*
63. SECOND LAW OF MOTION.—*The change of motion is always proportional to the impelling force.*
64. THIRD LAW OF MOTION.—*Action and re-action are always equal and contrary.*
65. AS REGARDS THE FIRST LAW OF MOTION, *That all bodies continue in a state of rest, or of uniform motion in a right line, unless by some impressed force they are made to change that state.*

If a body be impelled by two equal forces in contrary directions, the forces will destroy the effect of each other, and the body will remain at rest.

If the forces be unequal, the body will move in the direction of the strongest force.

When only one force acts on a body, the body obeys in the exact direction of that force.

66. AS REGARDS THE SECOND LAW OF MOTION, *That the alteration of motion is always proportional to the moving force impressed, and in the same direction in which that force is impressed.*

When in shooting at taw, if struck plump, as it is called, it moves forward exactly in the same line of direction in which it was struck; but, if struck sideways, it will move in an oblique direction, and its course will be a line situated between the direction of its former motion, and that of the force impressed.

When two or more forces, not in the same direction, act upon a body at the same time, as it cannot move two ways at once, it holds a middle course between the two directions. This is called the resolution of forces.

Thus, suppose a ship at sea, driven by the wind in the right line, A D, with such force as would carry it uniformly from A to D in one hour; then suppose the stream or current running in the direction A B, with such force as would carry the ship, through an equal space, from A to B, in one hour. By these two forces acting together at right angles to each other, the ship will describe the line A E C, in one hour; which line will be the diagonal of an exact square.

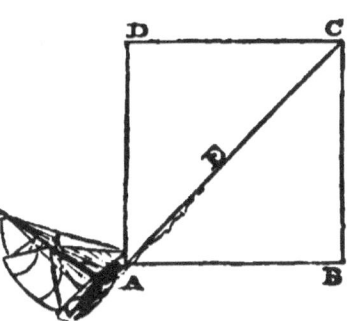

Other instances of this law:—A kite, acted upon by the wind and string; a fish striking the water with its tail; a boat rowed across the river, where there is a current.

67. AS REGARDS THE THIRD LAW OF MOTION.— *That action and re-action are equal and contrary.*

If we force our hand on the table directly downwards, the table presses upward; when a horse draws a load, the load, from its weight, acts as if drawing him back. We must not, however, think that this illustration of the law,—action and reaction are always equal,—means that the table really exerts a force upwards, because we press down upon it; it simply does not yield to your pressure, and may, therefore, be said to offer resistance.

68. By the axioms thus illustrated and applied "in abstracto" to every variety of motion, we establish a system of general doctrines concerning motions, according as they are simple or compounded, *accelerated, retarded, rectilineal, curvilineal,* in single bodies, or in systems of connected bodies; and we obtain corresponding characteristics and measures of *accelerating* or *retarding forces, centripetal* or *centrifugal, simple* or *compound.*

69. In the consideration of motion, there are several circumstances that must be attended to:—
1. The *force* which impresses the motion.
2. The *quantity of matter* in the body moved
3. The *velocity* and *direction* of the motion.
4. The *space* passed over in the moving body.

5. The *time* employed in going over the space.

6. The *force* with which it strikes *another body* that may be *opposed* to it.

70. EQUABLE MOTION is either simple or compound. *Simple* motion is that which is produced by the action or impressed force of one cause. *Compound* motion is that which is produced by two or more conspiring powers, that is, by powers whose directions are neither opposite nor co-incident, as illustrated in reference to the second law of motion.

71. ACCELERATED MOTION is that in which the velocity is continually increasing, from the continued action of the motive power. *Uniformly accelerated* motion is that in which the velocity increases *equally* in *equal times*.

72. MOTION is said to be *retarded* if its *velocity continually decreases*, and to be *uniformly retarded* if its velocity decreases equally in equal times.

73. MOMENTUM. The force with which a body moves, or which it would exert upon another body opposed to it, is always in *proportion to its velocity multiplied by its weight*, or quantity of matter; and this force is called the momentum of the body.[*]

74. For if two equal bodies move with different velocities, it is evident that their forces or momenta are as their velocities; and if two bodies move with the same velocity, their momenta are as their *quantities of matter*; therefore, in all cases their momenta must be as the products of their *quantities of matter and their velocities*.

75. Every boy who throws a ball or shoots a marble, is sensible that its force or momentum is in proportion to its velocity. The same marble will hit twice as *hard*, he will tell you, if it move twice as *fast*, or ten times as hard, if it move ten times as fast: substitute the word *momentum* for *hard*, and *velocity* for *fast*, and this principle of mechanics will be at once understood.

76. In every case of hard bodies striking each other, they may be regarded, for the sake of illustration, as compressing a very small strong spring between them.

77. When any elastic body, as a billiard-ball, strikes another

---

[*] Read Illustration No. 5.

larger than itself, and rebounds, it gives to that other not only all the motion which it originally possessed, this being done at the moment when it comes to rest, but an *additional quantity, equal to that with which it recoils*, owing to the equal action, in both directions, of the repulsion, or spring, which causes the recoil.

78. When the difference of size between the bodies is very great, the returning velocity of the smaller is nearly as great as its advancing motion was, and thus it gives a momentum to the body struck, nearly double of what it originally possessed itself.

79. This phenomenon is one of the most striking in physics, and seems to constitute the paradox of *an effect being greater than its cause;* and this has led persons, imperfectly acquainted with the subject, to seek from the principle a *perpetual motion*.

TIME.

80. TIME. The consideration of motion necessarily involves that of *time*, for no motion can be instantaneous. TIME is either *absolute* or *relative*.

81. ABSOLUTE TIME is a portion of *duration* whose quantity is only known by a *comparison with another portion*, and, consequently, the relation between any two parts of absolute time is not to be discovered.

82. RELATIVE TIME is a part of duration which elapses during any motion of a body, or any succession of external appearances.

83. To render time susceptible of mathematical discussion, it must be conceived as measurable, and to this end it is necessary to recur to some event which we imagine uniformly requires equal times for its accomplishment.

84. We are furnished with such an event in the complete *rotation of the earth upon its axis*, which works out a natural day as an apt and obvious unit of time.

85. This is divided into twenty-four parts called hours; each of these into sixty equal parts, called minutes; and each of these into sixty parts, called seconds. A second is the unit of time generally employed in *mechanics*.

CHAPTER VI.

ATTRACTION.

86. BY ATTRACTION is to be understood the tendency that all bodies have to approach each other.

87. *Attraction* is distinguished into various kinds; but as the causes of each are unknown, it is uncertain whether some of them be not different modifications of the same.

88. These various kinds of attraction are, 1. *The attraction of gravitation*, when acting at sensible distances, as in the case of the moon lifting the tides, &c. 2. Of *Cohesion*, when acting at very short distances, as in the substance of different bodies. 3. Of *Electricity*. 4. Of *Magnetism*. 5. Of *Chemical affinity*, when acting on fluids or gases, and, 6. *Capillary Attraction*, when acting between a liquid and the interior of a solid.

89. ATTRACTION OF GRAVITATION. It is one of the laws of nature discovered by Newton, and now universally received, that every particle of matter gravitates to every other particle; this law is the distinguishing feature of the Newtonian Philosophy. The attraction of gravitation, or gravity, is that force which causes all bodies near the earth to tend towards its centre with a force proportionate to their respective quantities of matter: it accordingly constitutes their weight.*

* Read Illustration No. 2, a, b, c, d.

ATTRACTION OF GRAVITATION.

90. It is by this power that descending bodies, on every side, fall on lines perpendicularly to the surface, and, consequently, on opposite sides they fall in opposite directions, and towards the centre, where the force of gravity is said to be accumulated; and by the same attraction bodies are kept on the earth's surface, so that they cannot fall from it.

91. All bodies that we know of, possess gravity, or weight. There is no such thing as perfect lightness. Smoke ascends only because it is lighter than the atmosphere, as cork swims, only because it is lighter than water. If the water be withdrawn, the cork sinks; if the air be withdrawn, as it is possible to be by the air-pump, the smoke will sink, like the cork.

92. Sir Isaac Newton demonstrated, that the attractive force of the earth decreases above its surface, in the same proportion that the square of the distance increases; and also, that the gravity of a body decreases below the earth's surface, in proportion to its distance from the centre.

93. Thus a body, which weighs on the earth's surface one pound, will, at two thousand miles from the centre, weigh but half a pound; at one thousand miles from the centre, but a quarter of a pound; and so on, until at the centre it loses all its weight.

94. The force of gravity, or general attraction, is such, at the surface of the earth, that in the first second of time it gives to a body, allowed to fall, a velocity of *sixteen* feet: *forty-eight* in the second; and *eighty* in the third. Knowing this law, we can easily tell the height of any high body, such as a tower or monument, by noting the time taken for a small body to fall. The annexed table shows the rate at which bodies fall; each of the triangular portions representing 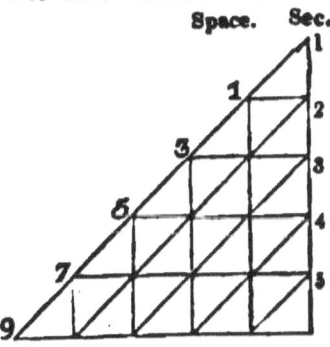 sixteen feet, the figures on the side the seconds.

95. In all places equi-distant from the centre of the globe, the force of gravity is equal. The earth, however, is not a perfect sphere, but a spheroid; having its equatorial larger

than its polar axis, by about thirty-seven miles; hence the force of gravity is less at the equator than at the poles, because the centrifugal force of the earth at the equator, diminishes the gravity: this is proved by the necessity of making the pendulum shorter at the equatorial than at the polar regions. Hence seconds pendulums, which, in the latitude of London, must be 39.13 inches, require to be one length shorter, or 39.12 at the equator.

₊ The attraction of electricity, magnetism, and chemical attraction, will be considered in these several subjects.

CHAPTER VII.

OF THE CENTRE OF GRAVITY.

96. THE centre of gravity in a body, is that point about which all its *other parts equally balance each other*. Most boys have often balanced a stick on the finger, or upon the chin. In performing this feat, they have only to keep the chin or finger exactly under the point, which is called the centre of gravity.

97. The method of finding the centre of gravity will depend entirely upon the form of the body. If in a cylindrical rod, it is only necessary to balance it, and the centre upon which it turns (we must here understand the centre of the rod itself, not the centre of its circumference) will, of course, be the centre of gravity; if of a circular flat body, it will be the centre of the circumference; if of a sphere, the centre of the sphere.

* Read Illustration No. 3.

98. The centre of gravity of a cylinder will be the same as that of the balanced rod, that of a cone at the point G, which is nearer the base. Thus it will be seen that broadness of base, and lowness of centre, are necessary to ensure stability.

Thus, if a line from the centre of gravity, on the following block E F, falls outside the base of the object, as in G H, the body will fall; if it falls on the base, G *a*, it will *totter*; it within the base, as at D *b*, or I *e* it will be secure.

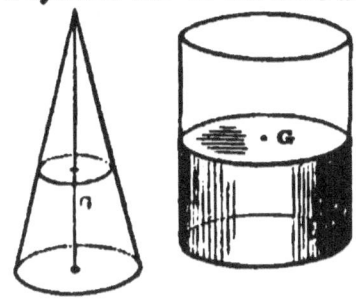

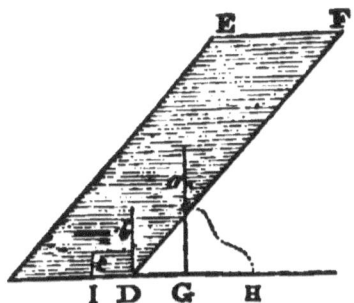

99. If we wish to find the centre of gravity in a board, or instance, of the following, or similar shape, represented by the figures *a*, *e*, *b*, *d*, let it be suspended from any point, as *a*, and the cord of a plummet, *a b*, be attached to the same point, the centre of gravity of the board must be somewhere in the direction of the plummet, (as may be seen in the cut,) and a chalk line, left on the board, must pass over the centre of gravity.*
If the board be then suspended at another point, at *d*, and another

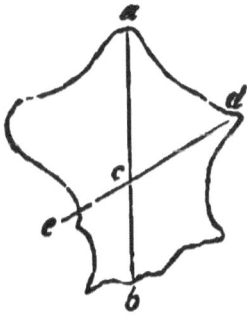

* Read Illustration No. 4.

chalk line, $d\ e$, be made in the same, the place c, where the two lines cut each other, will be the centre of gravity; and the board, when supported by a cord, attached there, will be equally balanced.

100. From want of attention to the centre of gravity, waggons and stage coaches, laden heavily on the top, will, upon coming on a road slightly inclined, immediately upset. If, as in the former case, a line from the centre falls within the edge of the wheel, which is here the extreme of the base, no danger need be apprehended; if without it, the waggon will fall over.

In rope-dancing it is necessary for the performer to have a long pole, loaded at its two ends. When he finds himself inclining too much in one direction, he throws out his pole in the opposite point, and thus recovers his balance, by bringing the centre of gravity vertically over the rope.

101. The *broader* the base and the *nearer* the line of direction is to the *middle* or centre of it, the more firmly does the body stand; on the contrary, the *narrower* the base, and the nearer the line of direction is to the *side of it*, the more easily may the body be overthrown.

102. If a plane be inclined on which a heavy body is placed, the body will *slide* down upon the plane while the line of direction falls within the base, *fig. a,* but it will *roll* down when that line falls without the base, *fig. b.*

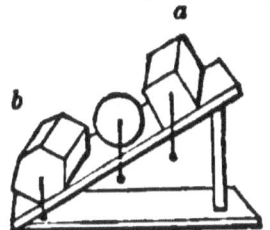

103. The centre of gravity in our bodies must constantly be adjusted, or we should be continually falling. When we walk up a hill we lean forward; when we walk down we lean backward; when we stand on one foot we lean sideways; in each case, that the centre of gravity may not overhang the base. When a man rises from a chair, he first bends his body considerably forward, and draws his feet backward. A man who carries a load on his head walks erect; he who carries a load on his back stoops; and he who carries it in his arms leans backward.

104. A man standing with his heels close to a perpendicular wall, cannot pick up any thing laid before him, because the wall prevents his throwing part of his body back, to counterbalance the head and arms, which must project forward. Many a wager has been won, and many a laugh raised, by setting thoughtless persons to do this.

CENTRAL FORCES.

105. Those actions or motions of bodies, which have reference to a fixed point, or centre, bear the common name of central forces. These forces again bear the name of CENTRIPETAL or CENTRIFUGAL.

106. The CENTRIPETAL force has a tendency to urge a body towards the centre, as in a whirlpool at sea, or a whirlwind on land. This force is exhibited in the cuts at the end of this chapter. It may also be called *gravity.*

107. The CENTRIFUGAL force has a tendency to throw a body from the centre, as the wet from a mop, when it is trundled; or the mud that flies from a wheel when it is turning.

108. One of the most practical illustrations of the centrifugal force is to be found in the action of mill-stones. The grain

is introduced into the centre of the stones, while the upper stone is rapidly whirling; the grain partakes of the motion, and is, by the centrifugal force, thrown out in the shape of flour, at the circumference.

109. Another practical illustration of the centrifugal force, 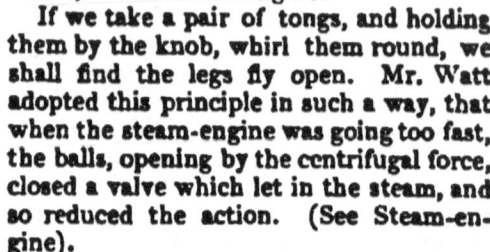 is seen in the governor-balls, as they are called, in the steam-engine.

If we take a pair of tongs, and holding them by the knob, whirl them round, we shall find the legs fly open. Mr. Watt adopted this principle in such a way, that when the steam-engine was going too fast, the balls, opening by the centrifugal force, closed a valve which let in the steam, and so reduced the action. (See Steam-engine).

110. By the CENTRIFUGAL force, the earth, instead of being a perfect sphere, is an oblate spheroid, bulging out seventeen miles at the equator, in consequence of its daily rotation. In the planets Jupiter and Saturn, whose rotation is much quicker, the equator bulges out still more.

Centripetal Motion.   Centrifugal Motion.

111. The knowledge of the action of the centrifugal force is of great practical use as regards health. If a person were to be laid on a wheel, with his head to the circumference, when the whole came in motion, the blood would fly towards his head, and occasion apoplexy. Thus the roundabouts of a country fair are injurious, while swings are comparatively innoxious. In one case the head is near, and in the other far from the centre, which makes the difference.

MECHANICS.

CHAPTER VIII.

OF THE MECHANICAL POWERS AND THEIR APPLICATION.

1. MECHANICS is the science of *motion* and of *moving powers*, as applied in the various engines called MACHINERY, employed in overcoming the resistance of *Inertia*.

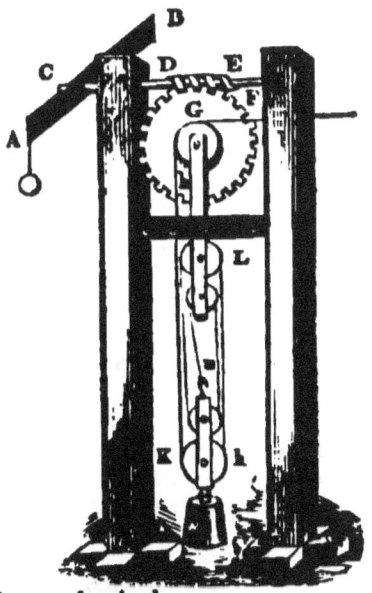

2. The MECHANICAL AGENTS are employed to *measure time*, to *move ships* and *carriages*, to *raise weights*, to *shape wood*, and *work metal*; to overcome the resistance of *air*, of *water*, and of *cohesion*. To draw out and form materials, and to combine them into new fabrics.

3. In the application of *mechanical agents* a combination of them frequently takes place. The instrument formed by this combination is called a *machine*. But all machines are composed of one or more simple powers called the *mechanical powers*.

4. The MECHANICAL POWERS are essentially but two in number, yet are usually considered as six, viz.

1. THE BALANCE AND LEVER 4. THE INCLINED PLANE
2. THE WHEEL AND AXLE 5. THE WEDGE
3. THE PULLEY 6. THE SCREW.

The three first are but assemblages of *levers*, and the three last but *inclined planes*, as will be hereafter explained.

C

26 MECHANICS.—THE LEVER.

5. OF THE LEVER. The *lever* is a bar resting on a support called a *fulcrum* or *prop*, for the purpose of raising, by a *power*

applied at one end, a *weight* at the other. In the cut above the slab is the *weight*, the small stone upon which the bar rests is the *fulcrum*, and the *hand* the *power*.

6. In the lever three things are to be attended to:—
    1st. The *power* which presses on the lever,
    2nd. The *weight* to be raised,
    3rd. The *situation* of the fulcrum.

7. The advantage results from having the *power* much further from the *fulcrum* than the *weight* is from it.

8. If the distance from the *power* to the fulcrum be five times greater than the distance from the *weight* to the fulcrum, a force of *one* pound in the *power* will balance *five* in the *weight*.

9. But here it is to be noticed that the end of the long arm of the lever will, as it turns on the prop, pass through a *space* five times greater than that of the short arm, and this brings us to the great axiom in mechanics, viz., *what is gained in power is lost in time.*

10. Levers are of three kinds: of the FIRST kind when the

fulcrum F is placed *between* the weight B and the power A.

Of the SECOND kind when the *weight* W is between the *fulcrum* F at one end, and the *power*, P, at the other.

Of the THIRD kind when the *fulcrum* F is at one end, the weight W at the other, and the *power*, P, between them.

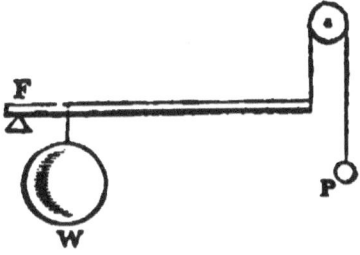

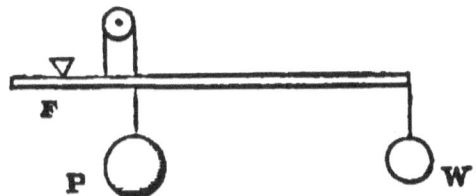

11. LEVERS OF THE FIRST KIND are to be seen in a variety of instruments, scissors, snuffers, pincers, which are made of two levers acting contrary to each other. The fulcrum or centre of motion being the pivot on which they move.

12. The common claw hammer; a man digging with a spade; and a poker raising coals in a grate, are levers of this description: in the latter the coals are the weight, the bar the fulcrum, and the hand the power. The common balance is also

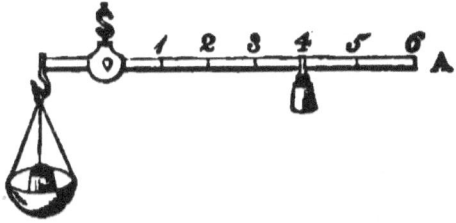

lever of the first kind, as is also the statera, or Roman steelyard. In the case of the balance, the arms of the beam must be equidistant from the fulcrum; in the steelyard, the longest arm is that to which the power is to be applied, (A).

The mechanical action of the steelyard is as follows. The body to be weighed is placed in the scale, and a sliding weight is placed on the beam. The scale and short arm exactly balance the long arm. If the weight being at *three* inches distance balance *six* ounces, at *four* it will balance *eight*, at five *ten*, and so on.

13. **LEVERS OF THE SECOND KIND.** This kind of lever is to be seen in the common wheelbarrow; here the *wheel* is the *fulcrum*, the resistance or weight is the load in the barrow, and the power the man who holds up the shafts.

The oars of a boat also present another instance; here the water is the fulcrum against which the blade of the oar

presses. This lever shows too why two boys carrying a burden upon a pole may bear unequal shares of weight, by shifting the burden *nearer to* or *farther from* each other.

Cutting knives, nut-cracks, a pair of bellows, horses drawing a plough, are levers of this kind.

THE TERTIARY LEVER.

14. The advantage gained in this kind of lever is as great as the distance of the *power* from the *prop* exceeds that of the weight.*

15. LEVERS OF THE THIRD KIND. To the third kind of lever, in which the power acts between the fulcrum and the resistance, may be referred the common tongs, shears for shearing sheep, and, above all, the limbs of animals. The human arm affords one of the most striking illustrations. Here the *prop*, or *fulcrum*, is the socket at the elbow, the *power* acts at a small distance from the socket, A, where the tendon, by the construction of the biceps muscle, C, draws up the arm, or lets it down.

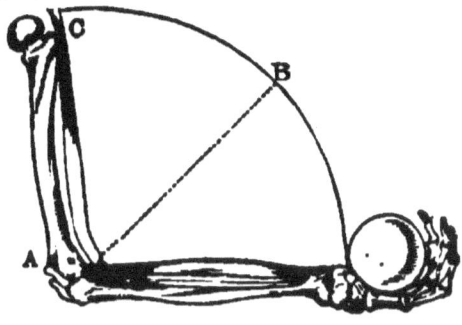

16. In all levers of this kind there is a *loss of power;* but, in the case of the human arm, the power is, generally, quite sufficient, as the hand is thus made to move quickly over a large space, which is much more useful to us, than to be able to move a large weight slowly over a small space. The muscle, by contracting its fibres less than an inch, raises the elbow 20 inches; and if it overcomes a force of 50 lbs. at the hand, it must be acting with a force of, at least, twenty times as

---

* To find what power will balance any given weight, state the following proportions:—
As the distance between the power and the fulcrum is to the distance between the weight and fulcrum, so is the weight to be raised to the power.
  Example:—When the power is 12 feet from the fulcrum, and weight 3 feet from it, what power will balance 1,200 lbs.?
  As 12 : 3 : : 1200 : 300 answer.

As the distance between the weight and fulcrum is to the distance between the power and prop, so is the weight.
  Example:—What weight, at 15 inches from the fulcrum, will be balanced by a power of 150 lbs., 75 inches from it?
  As 15 : 75 : : 150 : 750 answer.

intense, or of 1000 lbs., showing the extraordinary strength of muscle.

17. The wheels of a clock, and those of a watch, are levers of the third kind, the power being by a pinion near the centre, and the resistance acting upon the teeth of the circumference.

18. COMBINED LEVERS. Levers are combined in a great variety of ways. In the following cut we may obtain a slight notion of the principle upon which such combinations are made.

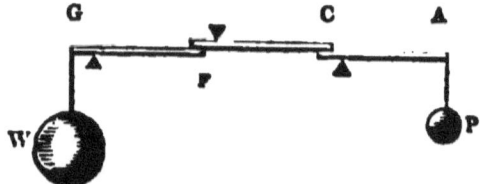

The power, P, brings down A, and thus raises C, bringing down F, which consequently raises G; and thus, if properly supported, will balance W.

CHAPTER IX.

OF THE WHEEL AND AXLE, AND THE PULLEY.

19. THE WHEEL AND AXLE is a machine much used, and is made in a great variety of forms; it consists of a wheel fixed to an axle or cylinder, so as to turn round together; sometimes it is a cylinder with projecting spokes, and the power being ap-

plied to the circumference of the wheel, as P, or to a han dle forming a larger circumference, as in the cut B, while the weight to be raised is fastened to a rope, which coils round the axle, or cylinder.

20. The *wheel* and *axle* may be considered as a kind of perpetual lever, of which the fulcrum is the centre of the axis, and the long and short arms, the diameter of the wheel and the diameter of the axis. From this it appears, that the larger the wheel and smaller the axis, the greater is the power of the machine; but then, as is common to every other mechanical contrivance, what is gained in power is lost in time—the weight must rise slower in proportion.  Sometimes the power is applied at the axis of the wheel instead of the circumference of the wheel, as in the case of coach or waggon wheels, and sometimes the wheel is a solid or a hollow cylinder, as in the subjoined cut, representing a roller.

21. It is proved by GEOMETRY,[*] that the circumferences of different circles bear the same proportion to each other as their respective diameters do; consequently the advantage gained by this mechanical power, is in proportion as the circumference or the diameter of the wheel is greater than that of the axis; hence the velocity of the power will be to that of the weight, as the circumference of the wheel is to that of the axis.

22. *Example:*—Suppose a water wheel to be 12 feet in diameter, and to turn an axle of 1 foot, the power acting at the circumference of the large wheel moves over 12 times the space which the circumference of the axle moves over; hence 12 cwt. may be raised by the power of 1 cwt.

The wheel and axle is, in principle, nothing else than a

---

[*] Bonnycastle.

lever, and has been called a perpetual lever. The subjoined diagram being a section of the wheel and axle, shows that the power and the weight act on each other through the intervention of the lever, A, C, B, whose fulcrum is at C, the axis of the machine, and the power, P, will balance the weight, W, in the same way as the power and weight of the common lever balance each other.

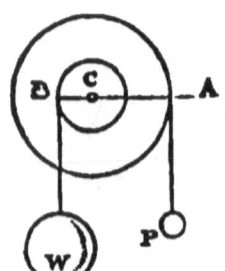

23. The wheel and axle is applied to a great variety of purposes. The common windlass, for raising water from a well, is one of its modifications. The capstan, used on board a ship, is an upright axle. The common winch, with which a *grindstone is turned*, or a crane worked, or a watch wound up.

24. That part of a common watch, called the fusee, affords not only a beautiful illustration of the principle of the wheel and axle, but of a very ingenious contrivance. To understand this, we must have a notion of the principle upon which watchwork acts.

25. The watch derives its motion, in the first case, from the principle of elasticity,—a property of matter already explained. The steel spring, called the main spring, A, is enclosed in the box, B, one end being fixed to the axis within the box, and the other to the circumference, the axis being immovable, and the box or barrel turning freely round it. The barrel, which thus contains the spring, is con-

nected, by means of a small chain, with the fusee, F; which.

MECHANISM OF A WATCH.

consists of a conical axle, in which is cut a spiral groove, to receive the chain as it is wound off the barrel.

26. Now, when the watch is wound up, which is done by placing the key in the axis of the fusee, and turning it in the direction, B, F, the barrel is made to revolve, and give off its chain to the fusee; while, at the same time, the main spring is coiled closely round the axis. We will suppose, that while the main spring is closely coiled up, that its power of elasticity is strongest in its effort to unbend itself; and, consequently, its action in giving rotation to the barrel is greatest.

27. But, if there were no means of equalizing this unequal action of the spring, we should have no uniform motion in the watch, and the first revolution of the hands would be performed in half an hour, and the last in two hours. To prevent this, as the force of the spring lessens, and the revolution of the barrel *weakens*; the radius, by which the chain acts upon the fusee, *increases*; and exactly as the spring becomes relaxed, it is pulling at a larger and larger part of the fusee, and so keeps up an equal effect on the general motion.

28. By means of a wheel, which is very large in proportion to the axle, forces of great intensities may be balanced; but, as very large wheels would be inconvenient, the end is obtained by using two or more wheels of a moderate size. In the engraving three wheels are seen, thus connected:—teeth on the *axle* of the smaller wheel acting on six times the number of teeth in the circumference of the second wheel, turns only once for every six times that the second turns; and this wheel, by turning six times, turns the third wheel once. The first wheel turns, therefore, thirty-six times, for one turn of the last.

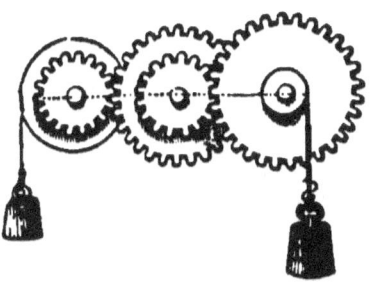

29. It is upon this principle that clocks and watches are made—a principle that involves the great axiom in all machinery, that what is gained in *power is lost in time*. By increasing the number of wheels, time-pieces are made which will go for a

year; and might, by still increasing their number, be made to go for a hundred, or a thousand years.

30. Wheels may be connected by bands as well as by teeth,

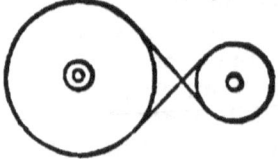

as seen in the common spinning-wheel, turning lathes, and in many others. A spinning-wheel, of 30 inches in circumference, turns a spindle of half an inch, 60 times for one of its own revolutions.

OF THE PULLEY.

31. The PULLEY is a small wheel turning on an axis, with a drawing rope passing over it, in a groove, to keep it from slipping. This wheel is usually fitted in a thick piece of wood, called a box, or block, so as to be movable round a pin passing through its centre. The wheel in this case is called the sheave.

32. Pulleys are of two kinds—*fixed* and *movable*. The fixed pulley gives no mechanical advantage, but is of great advantage in *altering the direction* in which it may be applied. The movable, on the contrary, *doubles the power*, which may be increased in any ratio whatever, by increasing the number of pulleys.

33. A simple pulley, supporting a weight of ten pounds, has

the weight divided on each end of the rope, and a man holding one end would bear only the weight of five pounds; but to raise the weight one foot, he must draw up two feet of rope; therefore, by the pulley, he lifts five pounds two feet, when, without it, he would have to lift ten pounds one foot.

34. The advantage of a pulley in changing the direction of a power, may be seen on board a ship to great advantage—a variety of ropes are pulled in a direction contrary to that in which the power acts. When the anchor is hoisted by the cathead, the pull is horizontal, but the mass of iron is raised vertically, and the same with many others. Without pulleys a ship could not be managed.

35. By means of a single fixed pulley, a man, being slung in the bight of a rope, may pull himself up to the beam upon which the pulley is fixed, by merely exerting a force equal to half his own weight; for, as in the case of the simple pulley and weight, just alluded to, the weight is divided—the part upon which he sits, and the part which he holds in his hands.

36. A man, by a pulley thus employed, may go down into a well, to save a fellow-creature who may have fallen into it; or, he might make his escape from a house on fire.

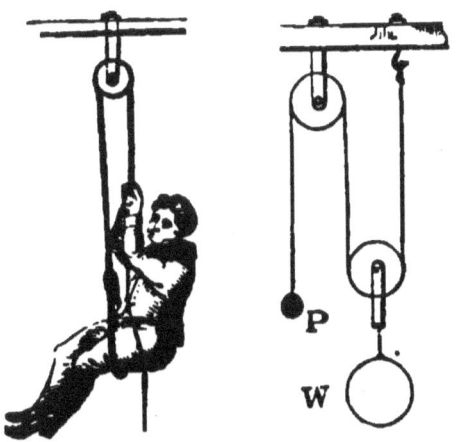

37. Pulleys are combined in a great number of ways; as in the diagram, the power of one will balance the resistance of two; for, as the weight, W, is supported by two strings, each of these will support the half of W; consequently, if the weight, P, be equal to half W, an equilibrium will take place.

38. The following diagrams represent two sets of pulleys, one with blocks, the other without them. Combinations; are however, the same in principle; there being two pulleys in each compartment, and the same rope passing over all the four, and the weight being thus supported by four folds of string, each of these supports one-fourth of the weight, W; therefore a force applied by the hand, equal to one-fourth of the weight, establishes an equilibrium.

THE PULLEY.—WHITE'S PULLEY.

In the use of the pulley, or a combination of pulleys, the advantage that may be gained is greatly diminished by the friction of the axles, and of the ropes. Too complex a combination, therefore, of pulleys, would not be of service, as the

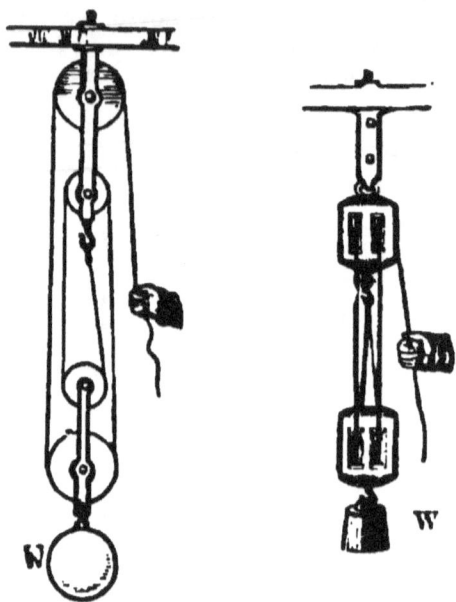

friction would not be increased without a proportional advantage; and, from the complexity of the machine, would be more liable to be put out of order.*

39. WHITE'S PULLEY.—In the system of pulleys already described, every pulley has a *separate* axle, and each pulley turning on its axle, must, consequently, be attended with *separate* friction. To obviate this defect an ingenious con-

---

* In general the advantage gained by pulleys is found by multiplying the number of pulleys in the lower block by 2.
A pair of blocks with a rope is called a *tackle*, as seen in the right-hand diagram.

trivance has been resorted to in White's patent pulley, by which all the pulleys on each block turn on the *same axis*, by which friction is reduced, and many other advantages obtained.

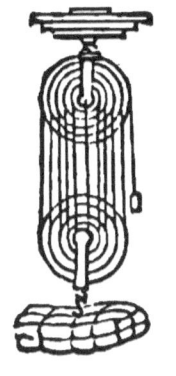

CHAPTER X.

THE INCLINED PLANE, WEDGE, AND SCREW.

40. THE next mechanical powers are the *inclined plane*, the *wedge*, and the *screw*. The two latter are only modifications of the former; so that, as before stated, there is only one principle involved in the action of these powers.

41. An *inclined plane*, is a smooth surface, raised higher at one end than the other, as is shown in the cut. The mechanical advantage arises from its supporting a portion of the weight laid upon it, the remainder being kept from sliding down the plane, by the power which is applied in a contrary direction.

D

THE INCLINED PLANE.

42. For example:—a force pushing a weight from *c* to *d*, only raises it through the perpendicular height, *e d*, by acting along the whole length of the plane, *c d*; and if the plane be twice as long as it is high, one pound at *b*, acting over the pulley, *d*, would balance two pounds at *a*, or any where on the plane; and so of all other proportions between the height and length of the plane.

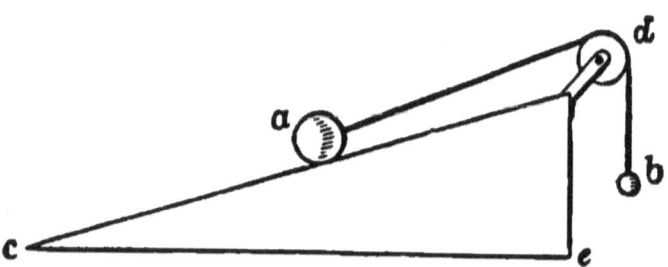

43. It has often struck travellers with astonishment, when viewing Stonehenge, how these enormous cross pieces of stone were raised to the elevation at which we see them, but the mechanical feat is by no means very wonderful: it would only be necessary to raise an inclined plane of earth in the direction

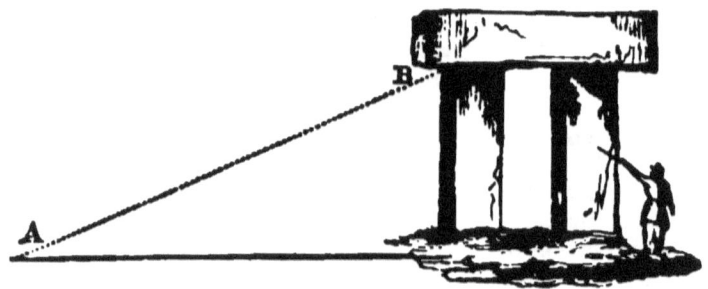

of the line, A B; and thus, by means of rollers placed under the stone, pass it to its situation on the top. The earth being then removed, the top stone would remain secure.

44. When a horse draws a load along a level road, all the

effort he requires to make, after the load is set in motion, is merely that which is necessary to counteract the effect of friction, and other resistances, arising from the inequalities of the road; but when a horse draws a load, where there is a rise of one foot in twenty, he is really lifting one-twentieth of the load, as well as overcoming the friction, and other resistance of the carriage.

45. A horse seems to understand the principle of decreasing the steepness of a hill, by extending the length of the road. Many horses will move up a hill in a zig-zag path, by which means they ease themselves considerably in their labour.

46. A flight of stairs are a set of inclined planes, modified to suit our convenience; and the less the height of the stair is, in comparison with its length, the easier it is to ascend it. The stairs leading to the stone gallery of St. Paul's Cathedral, are not more than two or three inches in depth, and may be ascended with amazing facility. A person stepping from one stair only, does not feel fatigued during his long ascent. Were the stairs the usual depth, he would be exhausted before he reached the top.

47. When a plane is inclined to the horizon one-third of its whole length, any body will be kept from rolling down that plane by a power equal to a third part of the weight of the body: if the height of the plane be equal to half its length, a power equal to half the weight of the body will support it.

48. No perpendicularly situated plane ought to come under the denomination of this article, because the plane in such a direction contributes nothing to the support or hindrance of the falling body, which descends with its whole force of gravity, unless prevented by a power equal to its whole weight.

49. It must be obvious, from the foregoing illustrations, that the *less the angle of elevation,* or the gentler the ascent is, the *greater will be the weight* which a given power can draw up; for the steeper the inclined plane is, the less does it support of the weight; and the greater the tendency which the weight has to roll, the more difficult will it be for the power to support it: hence the advantage gained by this mechanical power, is as *great as its length* ($c\ e$) exceeds its *perpendicular height* ($e\ d$).

THE WEDGE.—THE SCREW.

50. THE WEDGE.—The wedge consists of two inclined planes, joined together at the bases, as will be readily perceived by the figure; and the *advantage gained* is in *proportion as* the *length* of the *two sides* is *greater* than the *back*; that is, the sharper the wedge, the easier it will be driven. Technically, the power is increased by diminishing the angle, but the effect of the blow given is far greater than can be accounted for by any calculation of the effect of mechanical forces or powers.

51. The wedge is a mechanical power of singular efficacy, and the percussion by which its power is increased is precisely that force which we may with ease increase almost indefinitely.

52. By means of a wedge, the walls of houses may be propped, rocks split, ships raised up, and a resistance overcome, to which it would be impossible to apply the lever, wheel and axle, or pulley; and, even were it possible to apply them, their power would be insufficient to produce the effect required.

53. To the wedge may be reduced the axe, hatchet, chisel, (if chamfered on both sides,) spade, shovel, needles, knives, punches; in short, all instruments which terminate in an edge or point, and grow gradually thicker upwards.

54. The Author of Nature, whose works uniformly display the utmost wisdom in his design, has formed the beaks of such birds as obtain their food by pecking, wedge-shaped, to enable them to dig into the ground, and to break the shells of fruit. The whole figure of a bird also partakes of this shape, and is, thereby, adapted for cleaving through the air. Fish also exhibit the same form, which enables them to pass with ease through the water.

55. THE SCREW.—The screw is an inclined plane, working round a cylinder. To prove this, cut a piece of paper in the form of the inclined plane, *c e*, page 38, wind this round a pencil, and the plane will form a spiral round it, which will give the thread of the screw, as in the cut.

THE SCREW.

56. The screw, strictly speaking, consists of two parts, which work within the other, as may be seen in the screws that fasten the shutters. The spiral groove in the inside of the hole, is called the concave, or socket, or, technically, the *female screw*. The spiral projection round the screw, which goes into the hole, is called the thread of the screw; the *distance between* the threads is called the pace, or step; the *hollow* between the threads is called the groove, forge, or channel.

57. The screw must be assisted by a lever, or winch, to assist in turning it. The *advantage* gained by this mechanical power, is in proportion *as the circumference of the circle made by the lever, or winch*, is greater than the *interval*, or *distance, between the spirals*, or threads of the screw.

58. Thus, supposing the distance of the spirals to be half an inch, and the length of the winch 12 inches, the circle described by the handle of the winch, when the power acts, will be 76 inches nearly, or about 152 half inches, and consequently 152 times as great as the distance between the spirals; and, therefore, a power at the handle, whose intensity is equal to no more than a single pound, will balance 152 lbs. acting against the screw; and, as much additional force as is equal to overcome the friction, will raise the 152 lbs.: and the velocity of the power will be to the velocity of the weight, as 152 is to 1.

59. Hence it appears, that the longer the winch is, and the nearer the spirals are to each other, so much the greater is the force of the screw.

60. Screws are much used in presses of all kinds. It is a screw that draws together the iron jaws of a smith's vice, and the carpenter uses them in situations where nails could not be driven. A common cork-screw is the thread of a screw, without the cylinder.

61. A perpetual screw, is the name given where a screw acts on the teeth of a wheel, so as to produce a continued

volution of the wheel, as in the machine for illustrating the mechanical powers, page 25.

CHAPTER XI.

RESISTING MEDIUMS, THE FLY-WHEEL, FRICTION, &C.

The Fly-wheel.

62. THERE are *two kinds* of resistance to the motion of bodies; one arises from the resistance given to them by the opposition of the air or other bodies: thus, a man riding or

driving swiftly, will feel a current of air against his face, even though the atmosphere be perfectly calm at the time. A bullet from a gun, or a stone from a sling, produces a whistling sound, from the velocity of its motion through the air. A boat passing through the water, a wheel turning on its axle, a rope running through a pulley, all meet resistance by rubbing against the bodies with which they come in contact, and this rubbing is called friction; and, in all applications of the mechanical powers, one-third must be allowed to overcome the friction of surfaces.

63. To remedy, or correct this imperfection, many contrivances have been resorted to: oil and grease are applied to axles and pinions; and, where the parts work upon each other, they are made of hard polished metal, to present as little resistance to their surfaces as possible.

64. The *other species of resistance*, is owing to the *inertion or inactivity of matter*. If a heavy body descends in a resisting medium, as air, this motion will be accelerated, but not in the same proportion as in vacuo, for, in vacuo, every instant of time produces an equal acceleration, but, in the air, the retardations increase continually, because the resistance increases with the increasing density of the air as it approaches the earth's surface. Hence it is, that in *vacuo*, all bodies descend equally fast, but in *resisting mediums*, the heaviest fall fastest. (See Pneumatics.)

OF FRICTION.

65. FRICTION is of two kinds, that by *rubbing*, and that by *contact*. The one is represented by a locked waggon going down a hill; the other, by a wheel touching the ground in its usual motion. The friction between rolling bodies is much less than in those that drag. Hence, in some kinds of wheelwork the axle moves on small wheels, or rollers in the inner part of the nave; large metal balls on the same principle are used in moving large blocks of stone, and diminish the friction in the greatest possible degree.

66. Friction, by rubbing, being occasioned by a degree of cohesive attraction, between the touching surfaces, and the roughness or inequalities of those surfaces; and, as the ine-

qualities of the same surfaces, their projections and cavities more readily enter into each other when the material is the same on both the surfaces, the friction is greater between such than between pieces of different, or of heterogeneous substances with dissimilar grain. Hence, the joints of mathematical instruments are made of steel and brass, and the axletrees of steel.

67. The friction of one piece of iron, wood, brick, stone, &c., on another piece of the same substance has been measured, by using the second piece as an inclined plane, and then gradually lifting one end of it until the upper mass begins to slide. The inclination of the plane, just before the sliding commences is called the *angle of repose*. This angle is different for different substances; for metals, the force required to overcome the friction between small pieces of them, is equal to about a fourth of the weight of the moving piece; and for wood, it is about a half, but for large pieces, or great pressure, the friction is proportionably less.

68. The friction of the lever is very little, that of the wheel and axle is in proportion to the weight, velocity, and diameter of the axle: the smaller the diameter of the axle, the less will be the friction. The friction of pulleys is very great, on account of the smallness of their diameter in proportion to that of their axles. In the wedge and screw there is a great deal of friction; screws with sharp threads, have more friction than those with square threads, and endless screws have more than either. In the application of all mechanical powers, one third must be allowed to overcome the friction of the surface.

THE FLY-WHEEL.

69. To regulate the motions of machinery, fly-wheels are introduced. A fly-wheel is either a heavy wheel, or cross bar loaded with equal weights; this being made to revolve upon its axis, keeps up the force of the power (as seen in the diagram of the steam engine, page 42), and distributes it to all parts of the revolution. The effect of a fly-wheel may be easily determined, by turning a common mangle of Baker's construction. The pendulum of a clock, acts in a degree as a fly.

70. In applying force to impel machinery, the mere transmission of the effect of power is not always sufficient; it is frequently necessary that the motion of the machine at the

point of action, should be equable and unvarying, and not subject to sudden jolts and irregularities of motion to which compound machines are often liable. This want of uniformity in the working of a machine may be produced by three different causes.

1st. By an irregularity in the action of the power, or *first mover*, which propels the machinery.

2nd. By a want of uniformity in the resistance, or load upon the machine.

3rd. Because the machine, in the different positions which its parts assume during the motion, may transmit the impelling power to the working point, with different degrees of force.

71. These various irregularities are equalized by the simple intervention of the fly-wheel, which is a large heavy wheel or disc, balanced on its axis, and so connected with the machinery as to revolve rapidly with it, receiving its motion from the prime mover: it collects all the irregular motion of the machine in one accumulated power, which it afterwards maintains and distributes with an uniform motion.

72. In mechanics the use of the fly resembles a reservoir which collects the intermitting currents, and afterwards dispenses the water in a regular stream.

73. The power of the fly-wheel to resist acceleration, is proportional to the square of its diameter; and, therefore, by

Potter's Wheel.

sufficiently increasing its size and weight, we may be enabled to equalize the most inconstant and irregular motions in machinery.

74. The advantages of a fly-wheel are sensibly perceptible when a man acts upon a winch, in which case the action of the power is very unequal; its effect is least when the man pulls upwards from the height of his knee; and greatest when the handle is on the uppermost side of the axis, and the hands push it forwards in a horizontal direction. If a fly-wheel be placed on the axis of the winch, these unequal effects are equalized, and the force becomes uniform.

75. Besides the use of the fly in regulating the action of machinery, it has another property, which is the *accumulating or collecting together the successive exertions of a power*. Thus, if a small force be applied in giving rotation to a fly-wheel, and be continued until the wheel has acquired a very considerable degree of velocity, such a quantity of force would be at length accumulated in its circumference, as would overcome a resistance utterly disproportionate to the immediate action of the original force.

76. It would be very easy, in a few seconds, by the mere action of a man's arm, to give such a force to the circumference of a fly-wheel, as would transmit an impulse to a musket ball equal to that it receives from powder. The same principle explains the force with which a stone may be projected from a sling. The thong is swung several times round by the force of the arm, until a considerable portion of force is accumulated, and then it is projected with all the collected force.

77. Much of the efficacy of the fly depends on the position assigned to it in the machinery: if it be used as a regulator of force, it should be placed near the prime mover; but if, on the other hand, it be used as a magazine of power, it should be nearer to the working point. No general rules, however, can be given for its exact position. The accumulating power of the fly has led some persons into error in supposing that it adds force to the machine, beside what is received from the first mover. That this is not the case is very plain, considering the perfect inactivity of matter, and its incapability of possessing any force that is not received from some effective agent. On the contrary, the fly never retains all the power communicated to it by the first mover; for the resistance of the air and friction, robs it of some force.

78. There are several other modes of regulating the motion of complex machinery, such as the pendulum of a clock. The spring of a balance wheel, and the dead beat escapement of Graham. These are explained in the chapter on the application of machinery

Dead beat escapement.

CHAPTER XII.

THE PENDULUM.

79. A PENDULUM is a heavy body, hanging by a string or wire, which is movable on a centre, and each swing is called a vibration, or oscillation. The vibrations are produced by the falling of the weight to the lowest part of the circle, and by the force acquired by the fall.

80. Galileo was the first who observed, that all the vibrations of the same pendulum, whether great or small, are performed nearly in equal times; and the longer a pendulum, the slower are its vibrations, the squares of the sines being inversely as their lengths.

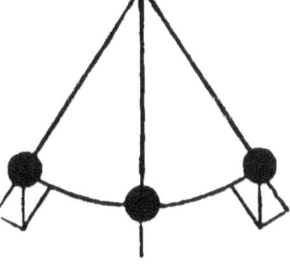

To show this, fasten any weight to the end of a string, and set it vibrating gradually; lengthen the string, and it will be seen that it requires more time. Also get two strings of unequal length, and produce a vibration as before, the shorter will vibrate faster than the other.

81 A pendulum vibrating seconds in the latitude of Lon-

don, must be 39.13 inches long; a pendulum vibrating *half* seconds, must be only a *quarter* of the length of one vibrating seconds. By the same rule, the length of a pendulum vibrating two seconds, must be four times as long as one vibrating seconds, or about 13 feet. The pendulum of the clock of St. Paul's is of this great length.

82. Heat expands, and consequently lengthens pendulums; and cold contracts, and shortens them. A pendulum, to vibrate seconds, must be shorter at the equator than at the poles, owing to the force of gravity, from which the pendulum derives its motion, increasing from the equator toward the poles; hence the length of the pendulum, which will swing seconds, must vary with the latitude.

83. As the smallest variation in the length of a pendulum makes a remarkable difference in time, many inventions have been made, to correct the contraction and expansion to which all other bodies are subject from the change of temperature. A pendulum, called the *gridiron pendulum*, is made of alternate bars of brass and steel, and so placed, that the expansion of the one corrects the contraction of the other, and thus the centre of oscillation is always kept in the same place.

84. The *centre of oscillation* is the point from which the *length* of the pendulum is estimated; it is in order to bring this point to the end of the rod that the ball is fixed. The time of a pendulum's variation is not affected by varying its weight, provided we do not shift the place of the centre of oscillation. To prove this, take two pieces of lead, one much heavier than the other, and fasten them to two pieces of string, of an equal length, and set them vibrating, they will perform their motions in exactly the same time.

The time of the vibration is not affected by the distance which it springs.

Set two pendulums in motion, as before, but make one commence swinging further than the other, they will both keep the same time.

85. As the length of a pendulum vibrating seconds remains universally the same, and may at any future period be ascertained, it has been considered as a proper basis for a system of uniform weights and measures. To determine its exact length is then necessary, as even the ten thousandth part of an inch

is of consequence. Captain Kater, by numerous and most ingenious experiments, has found it to be, in Portland Place, London, 39.13829 inches; and, reducing it to the level of the sea, 39.1393 inches. The French metre, the ten millionth part of a quadrant of the earth, was found to be 39.37079 inches in length.

The pendulum, by its application to clocks and watches, is become of immense importance.

EFFECTS OF TEMPERATURE ON THE PENDULUM.

86. If any substance could be found in nature sufficiently long for a pendulum that has not its dimensions enlarged or diminished by heat or cold, such substance would be most suitable for a simple attached pendulum, but all attempts to discover such substance have hitherto been ineffectual. Hence contrivances have been devised by ingenious men to counteract the effect of variable temperature in the pendulum, and some of them have succeeded in effecting this desirable purpose.

87. The wheel work of a common clock, it must be remembered, is merely employed to produce a given number of oscillations in the pendulum, and as the period of these oscillations depends on the length of the pendulous body; it will be obvious that one of the first essentials in the construction of a clock is to make the pendulum of a determinate length.

88. The length of a pendulum rod increases with heat; and it has been found by repeated experiments, that a brass rod equal in length to a seconds' pendulum, will expand or contract one thousandth part of an inch by a change of temperature of one degree of Fahrenheit's thermometer; and since the times of vibration are in a sub-duplicate ratio of the lengths of the pendulum, an expansion or contraction of one thousandth part of an inch will answer nearly to one second daily, so that if the clock be so adapted as to keep time when the thermometer is at 55°, it will lose ten seconds daily when the thermometer is at 65°, and gain as much when it is 45°.

89. The first of the inventions to remedy this defect in the measure of time is that by the celebrated Mr. George Graham, fig. 1. In this the rod or pendulum is a hollow tube in which a sufficient quantity of mercury is p^ut. This metal being easily affected by changes of temperature, kept as it rose or fell the centre of oscillation always equidistant from the point of suspension, as the mercury ascended while the rod of the pendulum descended or elongated. This pendulum performed with great accuracy, and its rate was determined by transits of fixed stars In the best form of the mercurial pendulum as now constructed, the bar is made of steel C D, and a glass vessel containing mercury O E is suspended underneath. The amount of compensation is determined by the quantity of mercury employed. When this pendulum is exposed to an increase of temperature the expansion of the rod C D and vessel O E will lower the centre of oscillation and thereby lengthen the pendulum while the expansion of the mercury will raise it, and supposing it to be at O thereby shorten the pendulum.

Fig. 1.

90. The gridiron pendulum already alluded to, is an ingenious contrivance first suggested by Graham, but invented by Harrison, a carpenter of Boston in Lincolnshire, in the year 1726. His rods of metal were placed in such parallel situations so as to resemble a gridiron, from whence its name. In the diagram the black lines are of one metal and the white spaces of another kind; the expansion of the white rods lengthens the pendulum, while that of the others will tend to shorten it. Since Harrison's time there have been various modifications of the compensating pendulum both in England and France, but none of them have been found to excel the prototypes of Graham and Harrison, which are all used one or other of them in the principal observatories.

Fig. 2.

HYDROSTATICS AND HYDRAULICS.

CHAPTER XIII.

HYDROSTATICS.

1. HYDROSTATICS treats of the weight and equilibrium of *non elastic* fluids at REST. When that equilibrium is destroyed *motion* ensues, and the science which considers the laws of such fluids in motion is called HYDRAULICS. (See page 62.)

2. Fluids are of two kinds, *non elastic* and *compressible.* Water, oil, and mercury are *non elastic.* Air, vapour, and gas are *elastic* and *compressible*, and fall under the branch of natural philosophy called PNEUMATICS. (See page 73).

3. PRESSURE of FLUIDS.—Solid bodies press downwards by the force of gravity, while their atoms or the particles of which they are composed by the attraction of cohesion are held together in a mass. Fluids on the contrary not only

press *downwards*, but also *upwards*, side ways, and in every direction *equally*. Thus it is that water *always finds its own level* as shown in the diagram, upon which principle fountains are constructed, and artesian wells are formed. The water being poured into the irregular vessel through the tubes A B C D E T will rise to the level line W W.

4. Fluids also press in *proportion to their perpendicular height*

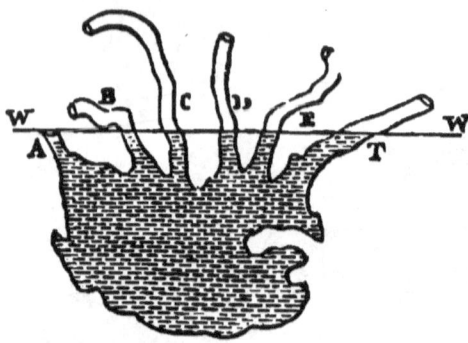

and the *area of the base of vessels containing them* without any regard to *their quantity*. If one vessel be in the shape of a *cylinder*, and the other in the shape of a cone —if the *bases* and the *perpendicular heights* be EQUAL, the *pressure* will be equal, although the solid contents and consequently the weight of the cone be only *one-third* that of the cylinder.

5. Thus in the vessel A the bottom *b c* does not sustain a pressure equal to the whole quantity of the fluid contained in the vessel; but only of a column whose base is B C, and height C E; also in the vessel F the bottom G H sustains a pressure equal to what it would if the vessel were as

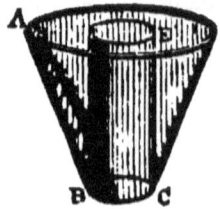 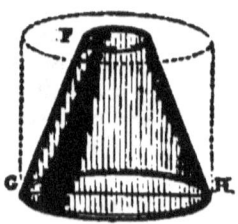

wide at top as at the bottom; hence the pressure on the bottom of a vessel may be *greater or less* than the *weight of water* it contains.

6. HYDROSTATICAL PARADOX. The above laws of pressure lead to what is called the *Hydrostatical Paradox*, which is of the greatest importance in this science. It is this. THAT A QUANTITY OF FLUID, HOWEVER SMALL, MAY BE MADE TO COUNTERPOISE THE LARGEST QUANTITY.

7. Thus if in a wide vessel *a b* made with a tube at its side *c d* communicating with the vessel, *water be poured* it will stand *at the same level* in *both* posts; consequently there is an *equilibrium* between them; and whatever may be the *shape* of the vessel, the *result* will be the same.

8. UPPER PRESSURE OF FLUIDS. The importance of this principle can be well illustrated by an instrument called the HYDROSTATIC BELLOWS. It consists of two pieces of wood joined together at the sides as in the common bellows. To this is attached a long metal tube A B which opens within. Water poured into this tube makes the upper board rise with great violence, and a weight from three to five cwts. may be sustained by it. A quarter of a pound of water in the tube will balance upwards of 300lbs., when the bellows are 16 inches long and 18 broad.

9. The explanation of this experiment is, that, if the tube have an area of the *fortieth of an inch*, and contains *half a pound* of water; this will produce a pressure of half a pound on every fortieth of an inch over all the interior of the bellows, from the effort which it makes to stand at *the same level* in the *bellows* that it does in the *tube*; and it would consequently support a weight, proportionate to a column of water equal to the *square of the base of the bellows multiplied by the height*.

10. HYDROSTATIC PRESS.—This is one of the most valuable hydrostatic machines yet constructed, and owes its invention to an application of the principle above described.

Instead, however, of a long column of water being made to produce the upward pressure. A B is a large solid plug or piston, which moves up and down in the main cylinder D, and fits it so as to be water tight. At the side of the piston is placed a FORCING PUMP, H, which is either worked by the hand or by a lever as in a common pump. The water raised by the pump from the reservoir G, is ejected by the pipe above E F, and being prevented from returning by the valve between G and H, is forced under the large piston B into the chamber of the cylinder.

∗ If the diameter of the forcing *pump* be one quarter of an inch, and the diameter of the *piston* B one foot, or 2304 times the force of H; then if a force of *one pound* be exerted on H it will force the water and transmit a pressure of *one pound* to *every part* of the force of the piston B which is *equal* to the force of the small piston H, producing a pressure of 2304 pounds on the bale or package C.

11. RULE FOR ESTIMATING THE PRESSURE OF FLUIDS ON HORIZONTAL SURFACES.—The general rule for estimating the pressure of fluids upon *horizontal* surfaces is *to multiply the height of the* FLUID *by the* EXTENT *of the* SURFACE *upon which it rests*. The product gives the bulk of the fluid, the *weight* of which is equal to the *pressure* upon the surface. Thus if the surface be *six* square feet, and the height of the fluid *three*, the pressure is equal to the weight of *eighteen cubic or solid feet* of the fluid. If it be water, a cubic foot of which weighs 62½ lbs. the pressure is equal to 1125 lbs.

12. PRESSURE ON SURFACES NOT HORIZONTAL.—In surfaces *not horizontal*, a different rule must be applied; for then, the pressure is *equal to the weight of a bulk of fluid, found by*

PRESSURE OF FLUIDS. 55

multiplying the extent of the surface on which it rests, into the depth of the centre of gravity of that surface, that is, of the point where being supported, the whole body remains balanced or at rest.

13. To find the pressure upon a *dam*, or the *flood-gate* of a lock, whether it be *upright* or *sloping* in the water, we have only to multiply HALF the *depth of the water* (which is the centre of gravity) into the superficial extent of the dam; this gives the *bulk of water, whose weight is the pressure on the dam.* If the water be *four* feet deep, and *twelve* broad, the dam, if perpendicular, is forty-eight square feet (the centre of gravity being at *half the depth,* or two feet), the pressure is equal to 96 cubic feet of water, or 6000 lbs.

Canal Lock.

14. The same rule extends to finding the pressure upon surfaces, either *plane* or *curved, horizontal, perpendicular,* or *slanting.* Thus if we would find the pressure upon the sloping slide of a pond; drop a line from the water to a *middle point* of the sloping side, between the water's edge and the bottom of the pond. Then *multiply the length of that portion of the line wetted, by the superficial extent of the side of the pond, covered with water;* the product will be the number of square feet of water on the side of the pond.

15. PRESSURE ON CYLINDERS, &c.—The pressure against the upright sides of a hollow tube, or cylinder, filled with water, such as a well or pipe, is found by multiplying the *curved surface** *under water* by *half* its depth *in* the water. Thus, if a pipe be 20 feet high, and its diameter, or bore, 4 feet, the calculation is 20 × 4 = 80 × $3\frac{3}{10}$ produces 252 square feet, which being multiplied by half the depth, viz., 10 × 252 = 2520 cubic feet of water, or above 70 tons pressure.

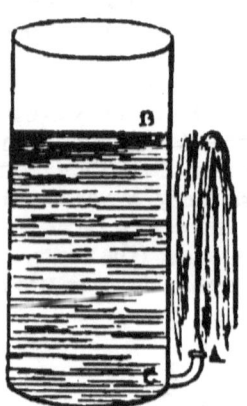

16. PRESSURE OF FLUIDS UPON SOLIDS IMMERSED IN THEM.—When a solid is immersed in a fluid, it sustains, upon the whole, an upward pressure equal to the weight of the fluid which it displaces. If, therefore, the solid be *equal* in weight to the water, it will be suspended *in* the water; if less than the fluid displaced, it will swim *at* or *partly above* the *surface*; if the weight of the solid be *greater* than the fluid, it will *sink*. Hence all bodies of *equal* bulks, which would *sink* in fluids, lose equal weight when *suspended* therein, and *unequal* bodies *lose in proportion to their bulks*.

SPIRIT LEVEL.—This instrument consists of a cylindrical glass tube filled with spirit of wine, except a small bubble of air at *b*. In whatever position the tube may be placed, the bubble of air will always lead to the highest part of it, when

placed in a perfectly horizontal position. Hence by this instrument we can easily ascertain whether a surface is level, or if two points are on the same level.

---

* The curved surface of a pipe or cylinder is found by multiplying 3.14159 or 3.320 into the product of its length, multiplied by its bore. The surface of a sphere, or globe, is in the same proportion, that is, as 3.320 is to the *square of its diameter*, that is the diameter *multiplied* by its *diameter*.

CHAPTER XIV.

SPECIFIC GRAVITY.

17. THE SPECIFIC GRAVITY of a body is its *relative weight* to *other bodies* of the *same bulk*, or of some particular body selected as a standard.

18. This STANDARD is *distilled water*, which at the temperature of $62°$ Fahrenheit, contains exactly 1000 *ounces* avoirdupoise in a *cubic foot*. When, therefore, we are told the *specific gravity* of a body, we know also the *weight of a cubic foot*, or 1728 inches, and may hence calculate the weight of any quantity of matter by simple *measurement*.

19. The method of ascertaining the *specific gravities* of bodies was discovered by ARCHIMEDES, in the following manner:

20. HIERO, king of Syracuse, having given a quantity of pure gold to a workman to make into a crown, suspected that the artist kept part of the gold, and alloyed the crown with a baser metal. He applied to ARCHIMEDES to discover the fraud. The philosopher having long studied it in vain, accidentally hit upon a method of verifying the king's suspicion. Going one day into a bath, he took notice that the water rose in the bath, and immediately reflected, that any body of an equal bulk with himself, would have raised the water just as much; though a body of equal weight, but not of equal bulk, would not raise it so much. From this idea, he conceived a mode of finding out what he so much wished, and was so transported with joy, that he leaped out of the bath, crying out, " I have found it, I have found it !"

21. Since gold was the heaviest of all metals known to ARCHIMEDES, it occurred to him that it must be of less bulk, according to its weight, than any other metal He, therefore, desired that a mass of pure gold, equally heavy with the crown when weighed in air, should be weighed against it in water, conjecturing that if the crown was not alloyed, it would coun

terpoise the mass of gold when they were both immersed in water, as well as it did when they were weighed in air. But upon making the trial, he found that the mass of gold weighed much heavier in water than the crown did; nor was this all—when the mass and crown were immersed separately in one vessel of water, the crown raised the water much higher than the mass did, which showed it to be alloyed with some lighter metal that increased its bulk.

22. METHOD OF ASCERTAINING THE SPECIFIC GRAVITY OF SOLIDS.—

The instrument used for ascertaining the specific gravity of *solid bodies*, is called the HYDROSTATIC BALANCE. It differs but little from the common balance, excepting that a hook is attached to the bottom of one of its scales, by which the substance to be weighed in water, may be suspended by a horsehair in the fluid, without wetting the scales. In these scales the body is *first weighed in air;* then, afterwards being suspended by a horsehair, it is weighed in *water*. The weight in *water* is now subtracted from the weight in *air*, and the weight in air is divided by the *difference* thus obtained.

23. Example :—If a piece of stone weighs in air 560 grains, and in water 360 grains. The specific gravity is obtained as follows :

    Gr.
    560 weighed in air
    360 weighed in water
    ———
    200 difference.

Now 560 divided by 200, gives 2.8, and if we reckon water at 1000, the specific gravity of the stone is 2800.

24. Or, proportionally, suppose a body weighs 1250 grains in air, and 830 in water, then 1250—830=420 grains, and as

420 : 1250 :: 1000 : 2976 the specific gravity of the solid in question.

25. GRAVITY OF SOLIDS LIGHTER THAN WATER — Some bodies, such as wood and cork, being lighter than water, a different method must be employed.

Pincers must be provided to keep the substance under water. First, the body is weighed in air; the pincers are then balanced under water, and affixed to the body to be weighed, which, being lighter than water, will raise the pincers, and cause the other scale to preponderate. The loss of weight being noted, the calculation is made as above.

26. Another method is to weigh the substance, say wood, in *air*, then to weigh in *water* a piece of metal sufficient to sink it. This is then fastened to the wood, and both are *weighed together in water*, when it is easily seen how much the wood weighs more than the water, and the proportion may be stated.

27. Example:—Suppose a piece of wood weighs in air, 16 ounces, and a piece of lead weighed in water, 12 ounces, and that when both are fastened together and suspended by a hair in water, the weight in water is only eight ounces, the difference, therefore, would be 4 ounces, and the wood was 4 ounces less heavy than *an equal bulk of water*; an equal bulk of water would thus have 20 ounces, and we may state

As 20 : 1000 :: 16 : 800

28. The SPECIFIC GRAVITIES OF FLUIDS are ascertained by the same principle. If a substance be weighed in two fluids, the weight which it *loses* in each is as the *specific gravity* of that fluid. Thus a cubic inch of lead loses 253 grains when weighed in water, and only 209 grains when weighed in rectified spirit. Therefore, a cubic inch of rectified spirit weighs 209 grains, an equal bulk of water 253. Hence the specific gravity of water is about a fourth greater than that of the spirit.

29. THE HYDROMETER.—The specific gravity of fluids is also found by an instrument called the HYDROMETER, the use of which depends upon the following propositions:

1. The HYDROMETER will sink in different fluids in an *inverse proportion* to the *density* of the fluids.

60   THE HYDROMETER.

2. The weight required to sink a *hydrometer* equally far in different fluids, will be *directly* as the *densities* of the fluids.

29. Each of these two propositions give rise to a different kind of hydrometer, the first, 1, with a graduated scale, the second, 2, with weights.

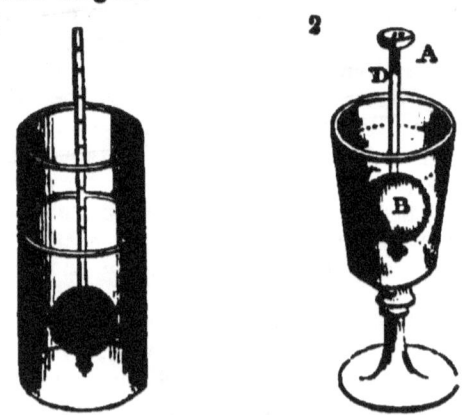

30. The first instrument is simply a glass or copper ball, with a stem affixed, on which is marked a scale of equal parts, or degrees. The point to which the stem sinks in any liquid, being ascertained, we can tell how many degrees any other liquid is heavier or lighter, by observing the point to which the stem sinks into it.

31. The instrument, No. 2. consists of a hollow copper ball having a steel stem D, of one-fortieth of an inch in diameter, supporting a small dish A at the upper end; at C just below the ball is a smaller dish, and the whole instrument is of such a weight that with the addition of 1000 grains in the dish A, it will just sink to the middle of the stem D, in distilled water at 60° of Fahrenheit.

32. If this *hydrometer* be immersed in any liquid whose specific gravity is to be determined, and weights are put in the dish A, until the instrument again sinks to the middle of the stem D, if the weights should be 500 grains, this added to the known weight of the instrument, 3000; the weight of the fluid displaced, will be 3500 grains.

We have now obtained the weights of equal volumes of

AERIFORM SUBSTANCES. 61

water and the other fluid, and hence we proceed as before, viz.: as 4000 is to 3500 : : 1000 to 875 the specific gravity of the fluid in question.

33. USE OF THE HYDROMETER.—The principal use of the hydrometer is to ascertain the specific gravity of distilled liquids, such as rum, brandy, or gin, and consequently their strength. But it is applicable to all fluids, and so great is the delicacy of this hydrometer that the difference in the specific gravity of one part in 40,000 can be detected.

34. There is another method of finding the *specific gravity* of a liquid, namely, by means of a bottle or flask made to hold exactly 1000 grains of distilled water. The bottle is filled with the liquid, then weighed; and from the *whole weight* is subtracted the *weight* of the *bottle*, the *remainder of course* is the *weight* of the *liquid* that fills it.

35. AERIFORM SUBSTANCES.—The *specific gravities* of aeriform substances are found much in the same manner. A glass flask of known size and furnished with a stop cock, is first weighed when emptied by the air pump; and afterwards when filled successively with *water* and with different *airs* or *gases*. Comparison of the weights give the specific gravities already described. (See Pneumatics).

When the specific gravity of any body is known, it is easy to calculate the weight of any given bulk of it, as the figures which denote the specific gravity, also denote the number of ounces avoirdupois in a square foot. (See table of specific gravities).

Example 1. What is the weight of 7 solid feet of lead?

11352 sp. gr. of lead.
7
―――――
16) 79464 ounces.
―――――
28) 4966 lbs. 8oz.
―――――
4) 177 qrs. 10 lbs. 8oz.
―――――
Ans. 44 cwts. 1 qr. 10 lbs.

Example 2. What is the weight of 48 solid feet of oak?

925 spec. grav. of oak.
48
―――――
16) 44400
―――――
28) 2775
―――――
4) 99 3 qrs.
―――――
Ans 24 cwts. 3 qrs.

HYDRAULICS.

CHAPTER XV.

1. The science of hydraulics comprehends the *laws* which regulate NON ELASTIC fluids when in motion, and the *action* of various machinery employed in connection with their motions.

2. Water can only be set in motion by *two* causes, 1st. the increased or decreased *pressure* of the *atmosphere*, which causes it to rise *above its own level* ; 2dly. by its *own gravity*, which induces it always to seek the *lowest level*.

3. All fluids which have a communication will be perfectly *level in the surface when at rest*, as already explained at page 51. Thus if two casks be united together by a pipe, going from one to the other, and if water be poured in, it will rise to exactly the same level in both. (See Hydrostatics).

4. From this law of hydraulics the most important effects arise which add greatly to the comforts of civilized life. Water from a reservoir may be distributed in pipes over a town and will rise to any height not exceeding the level of the surface of the water in the reservoir; hence it may be raised to the first or second floor above the street or even higher. The reservoirs of the new river company at Islington, and at the end of Tottenham Court Road, having the water at a higher elevation than many parts of London, can easily distribute it

to various parts of that city. Sometimes a high level is obtained by means of *forcing engines.*

5. FOUNTAINS.—As water will rise through bended pipes to the same level as the reservoir from which it proceeds, it enables us to form jets or fountains. If for instance a body of water be collected in a reservoir on the upper part of a house, and a tube or pipe descending from it to the garden be made to turn upwards having a very small bore, the water will rise in a jet and spout up *nearly as high* as the surface of the water in the reservoir. The water never rises the whole height on account of the friction of the pipe, the gravity of the water and the resistance of the air. If within the centre of any figure as that in the cut, a hollow ball be enclosed, it may be sustained on the top of a fountain or jet d'eau and made to dance upon its apex. An egg may be made to play in the same manner upon the top of a column of vapour, as seen also in the diagram.

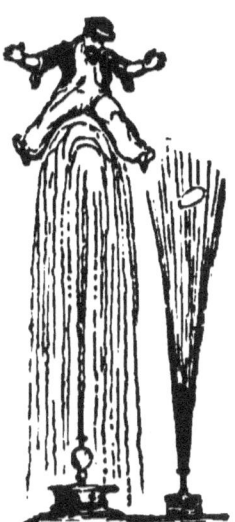

6. HORIZONTAL FORCE.—The HORIZONTAL DISTANCE to which a fluid will spout from a horizontal aperture in any part of the side of an upright vessel, below the surface of the fluid, *is equal to twice the length of the perpendicular to the side of the vessel, drawn from the mouth of the pipe to a semicircle described upon the altitude of the fluid;* and therefore the spout will be to the greatest distance possible from a pipe whose mouth is at the centre of the semicircle; because a perpendicular to its diameter, supposed parallel to the side of the vessel drawn from that point, is the largest that can possibly be drawn from any part of the diameter to the circumference of the circle.

7. Thus if the same vessel A B be full of water the horizontal pipe D in the middle of the side and the semicircle O E C, be described upon D as a centre with the radius D C or D O, the perpendicular D to be the diameter C D O is the largest that can be drawn from any part of the diameter to the cir-

cumference; and if the vessel be kept full the jet will spout

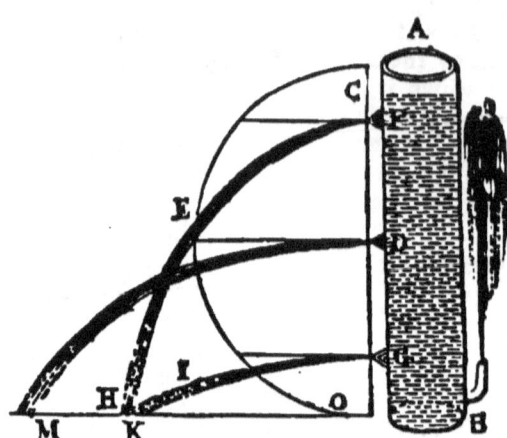

from the pipe C to the horizontal distance M which is double the length of the perpendicular D E. If two other pipes at F and G be fixed into the sides of the vessel at equal distances above and below the pipe D, the perpendicular F H, and G I, from these pipes to the semicircle will be equal; and the jets spouting from them will each go to the horizontal distance N K which is double the length of either of the perpendiculars F H or G I. The curves described by the spouting fluid in all the different situations will be of a parabola, being acted upon by the combined forces of the lateral pressure of the fluid in the vessel and the force of gravity.

8. The velocity with which water would issue from a given depth below the horizontal surface as at the aperture B, is the same as that which a heavy body would acquire in falling through a height equal to B C. The square root of the intermediate space being the guide. For if a small jet tube as at B be introduced in the side of the vessel A, the water will, making a small allowance for friction, and resistance of the air, rise to the level of the surface of the liquid at C.

9. RATIO OF PRESSURE. In order to make *double* the quantity of fluid run through one hole as through another of the same size, it will require *four* times the pressure of the other, and therefore the aperture must be four times the depth of the other below the surface of the water; and for the same reason, *three* times the quantity running in an equal time through the same sort of hole, must run with *three* times the velocity, which will require *nine* times the pressure, and con-

sequently the hole must be nine times as deep below the surface of the fluid.

If two pipes of equal bores be fixed in the side of a vessel, one four times as deep below the surface of the water as the other, a cup that holds a quart will be filled by the bottom spout in the same time that a pint is by the upper one.

10. EXPERIMENTS.—We may fill with water a cask in which we have made a hole, secured by a cork, in the lower part, and mark at the side by means of any scale or measure the height of 1 inch, of 4 inches, of 9 inches, of 16 inches, and of 25 inches. If we now take out the cork, and notice while the water is running out how long it takes to sink to 16 inches, it will require exactly the same time to sink from 16 to 9 inches, also exactly the same time to sink from 9 to 4 inches, the same time also to sink from 4 to 1 inch, and then the same space of time for the remaining water to run out.

11. THE SYPHON.—The syphon is a bent tube having one leg shorter than the other. It acts by the pressure of the atmosphere being removed from the surface of a fluid, which enables it to rise above its common level. It is a very useful instrument and is employed for the purpose of emptying liquors from casks, &c.

12. ACTION OF THE SYPHON. In order to make a syphon act, it is necessary first to fill both legs quite full of the fluid, and then the shorter leg must be placed in the vessel to be emptied. Immediately upon withdrawing the finger from the longer leg, the liquor will flow. If the perpendicular height of a syphon, from the surface of the water to its bended top, be more than 33 feet, it will draw no water, even though the other leg were much longer, and the syphon quite emptied of air; because the weight of a column of water 33 feet high, is equal to the weight of a column of air reaching from the surface of the earth to the top of the atmosphere. Mercury may be drawn through a syphon, in the same manner as water; but then the utmost height of the syphon must always be less

than 30 inches, as mercury is nearly fourteen times heavier than water. The syphon may be filled, by pouring some of the fluid into it, or by placing the shorter leg in the vessel, and sucking the liquor through a side tube. Some are made with a sucking-pipe attached to the longer leg. Syphons are extremely convenient for decanting liquors of various kinds, as they do not disturb the sediment.

13. INTERMITTING SPRINGS.—Upon the principle of the syphon also, we may easily account for *intermitting* or *reciprocating springs*.

Illust. Let A be part of a hill, within which there is a cavity E, and from this cavity a vein, or channel, running in the direction H B C. The rain that falls upon the side of the hill, will sink and strain through the small pores and crevices in the hill $d\ d\ d$, and fill the cavity E P with water. When the water rises to the level of O H, the vein H K will be full, and the water will run through it as a syphon, and will empty the cavity E P by degrees. It must then stop, and when the cavity is again filled, it will begin to run again.

CHAPTER XVI.

PUMPS.

14. THE PUMP.—The pump is at once the most common and most useful of all hydraulic machines. It was invented by Ctesebes a mathematician of Alexandria about 120 years be-

fore Christ. When the pressure of the air became afterwards known; (see Pneumatics) it was very much improved and is now brought to a great degree of perfection.

15. Of this machine there are three kinds, viz., the SUCK-ING, the LIFTING and the FORCING pump. By the last two water may be raised to any height with an adequate apparatus and sufficient power. By the sucking pump it can only be raised 33 feet above the surface. (See Pneumatics.)

16. THE SUCKING PUMP.—The *sucking pump* is an engine both with pneumatic and hydraulic. It consists of a pipe open at both ends, in which is a moveable cylinder or piston, as big as the bore of the pipe in that part wherein it works; and contrived by leathers or other means to fit the bore exactly, so as not to allow any air to pass between it and the sides of the pipe where it acts. In the piston of the common sucking-pump, there is a valve opening upwards, like a trap-door, to allow the air and water readily to ascend through it, but to prevent either of them from descending. This piston is called the bucket. It is moved up and down in the pipe by a rod fastened to a handle or lever, or such parts of the machinery as are intended to work it. The pipe usually consists of two parts, of which the first and wider part is called the barrel, or the working-barrel, because it contains the piston; and the other is called the suction-pipe, being of a smaller diameter.

At the joining of the working-barrel with the suction-pipe, there is a fixed valve opening also upwards. Lastly, the lower end of the suction-pipe is immersed in water, which is admitted into it through small holes to prevent the entrance of dirt; at the top of the working-barrel is a wide head, and a pipe for the delivery of the water which is raised.

17. The mode of raising water by means of a pump is this: when the handle is raised the piston rod descends, and brings the piston valve (called the sucker or bucket) to another valve, which is fixed, and opens inwards upon the piston; when the handle is drawn down, the piston is raised, and as it is air tight, a vacuum is produced between the two valves; the air in the barrel of the pump, betwixt the lower valve and the water, then forces open the lower valve, and rushes through to fill up this vacuum, and the air in the pump being thereby

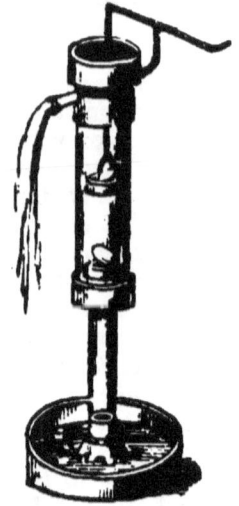

less dense than the external atmosphere, the water is forced a short way up the barrel. When the piston again descends to the lower valve, the air between them is again forced out by forcing open the upper valve, and when the piston is raised a vacuum is again produced, and the air below the lower valve rushes up, and the water in consequence is again raised a little farther. This operation continues until the water rises above the lower valve; at every stroke afterwards the water passes through the valve of the descending piston, and is raised by it on its ascent, until it issues out at the spout.

18. THE LIFTING PUMP.—In the lifting pump there is always a column of water lifted above the base is equal to the top of the piston, and whose height is equal to the distance from the piston to the head. It is evident that this weight will not be made less by diminishing the diameter of the barrel above the piston because fluids given in proportion to their bases and perpendicular altitudes. This pump is much used in great water works it is the simplest of all in its operations. (See Chain Pump).

19. THE FORCING PUMP.—The forcing pump consists of two parts or barrels, one similar to the common pump, and the other rising by its side, and connected to it near the lower valve. The water is first raised in the former part in the same manner as in the common pump, excepting, that the piston has no opening valve, but is solid, and the air is forced out through the side valve into the adjoining tube, or barrel. Through this valve, the water is also forced,

and the pressure of the descending piston makes it rise and enter the ascending tube and issue out at the top. The length of this tube may be increased at pleasure, provided a sufficient power be employed to force the water up.—It is hence the *forcing pump*.

21. In order to increase the force with which the water rushes out, there is a vessel surrounding this tube which encloses a volume of air. When the water rises, the air is compressed, and, as it is very elastic, it resists this compression, and the more so as the water is actively pumped up, and increases the compression; and hence the water is made to fly out with great violence.

22. The engines, employed with so much advantage in extinguishing fires, and which throw the water to a great height, are an example of the application of the forcing-pump.

CHAPTER XVII.

HYDRAULIC MACHINERY.

23. The earlier hydraulic machines appear to have been constructed upon the most simple principles, probably the first process towards the raising of water in a bucket was by means of a lever as here represented; (which is still the method in uncivilized nations,) or by the hands, or by a windlass as in our common draw well.

24. The PERSIAN WHEEL is one of the earliest engines used for raising water to a certain height; for which it is necessary that the diameter of the wheel should be somewhat greater than the level to which the water is intended to be received. It consists of a wooden wheel which revolves upon an axis or gudgeon; upon its circumference a number of buckets are suspended P, by strong pins, m such a manner that they become filled with water from the stream or reservoir, and hang upright as the wheel revolves, until they reach the top when they strike against the fixed troughs and are tilted, their contents being discharged into the trough, from whence it may be conveyed by pipes to the place for which it may be designed.

25. CHAIN PUMP.—The chain pump usually consists of a succession of long links of metal rods revolving like an endless rope over two wheels, $e\ f$, one of which f must be under water: on this chain between each joint is fixed a flat piece of wood or metal $d\ d$, usually square, which is supported and kept in its place by the projecting arms of the wheels f and e, though at the same time they are permitted to turn with the same freedom as the chain. The wheel e is turned by a winch which causes the whole chain to move, one side of it passes upwards while the other side is continually descending in the same direction. The ascending side of the chain is made to pass through a box or pipe, one end of which is immersed in the water, the other end reaching the upper wheel; this box corresponds in shape

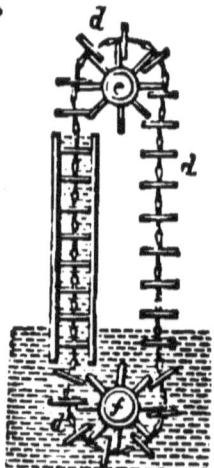

CHAIN PUMP.—SCREW PUMP.

with the size of the plates, which fit pretty closely and form the pump, the eight plates as they pass upwards through the pump forms a succession of cavities which are filled with water and constantly discharged at the top.

25. ARCHIMEDES' SCREW or water snail, as it is sometimes called, deserves also consideration not only for its antiquity, but for its usefulness in raising water. It is a metallic tube bent round a cylinder turning on an axis as seen in the cut. The lower end of the screw being inserted in the water A which is to be raised, when the handle at H is turned, a portion of water forces through the open mouth and is successively elevated through the various coils C, until it arrives at the top of the tube where it is discharged.

26. CENTRIFUGAL PUMP.—This is a valuable machine for raising water by means of the centrifugal force combined with the pressure of the atmosphere. The figure represents this machine. $g h$ is an upright spindle, so fixed that a rapid rotatory motion may be communicated to it by the winch $i: k$ m represents any number of curved pipes; so disposed and fixed to the spindle that their lower ends may be near it and be covered by the water to be received: and their upper ends which are quite open, to be extended to a considerable distance from the centre of motion, and bent downwards from the top to prevent the scattering of the water. Upon putting the

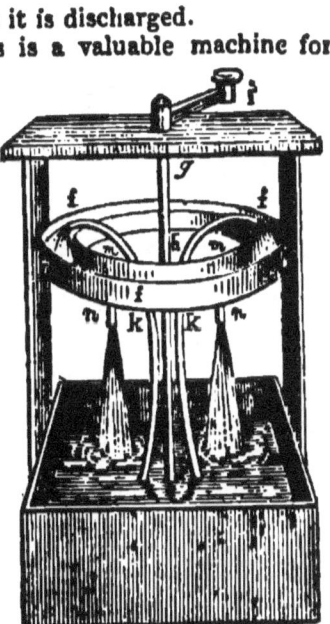

machine in action the several pipes must be filled with water which will be retained in them by a valve opening upwards, and placed near the bottom of each pipe. The machine is then put in motion by turning the winch. The higher ends of the pipe *m* will now describe a much larger circle than the ends below; and consequently, such a centrifugal force or tendency to fly off and empty the pipes, will be introduced at the upper end as will produce a vacuum capable of raising

a column of water through the pipes; *fff* is a circular pan to receive the water as it runs from the upper ends of the pipes, and *n n* are spouts by which the water runs off.

27. BAKER's MILL.—This machine owes its efficacy to the centrifugal force. It consists of a long cylindrical pipe, having a funnel at A, and terminating in a pivot, turning in a socket at B. About A is an axis E passing through a frame, and carrying with it the upper millstone. At the bottom of the pipe at B is a cross pipe C D at the opposite sides of which are two apertures from which the water poured into the funnel at A, spouts with considerable velocity, and merely from the resistance of the air gives motion to the machine.

28. WATER MILLS are an ancient great means of power; they are called *undershot* when carried round by a current in which the floats dip, and *overshot* when the water from above falls on the floats.

PNEUMATICS.

CHAPTER XVII.

DEFINITION—THE AIR—HEIGHT OF THE ATMOSPHERE—AIR A MATERIAL BODY—COLOUR OF THE AIR—ITS WEIGHT—PRESSURE, &C.—AGGREGATE PRESSURE OF THE AIR ON THE GLOBE.

1. PNEUMATICS is that branch of the physical sciences which treats of the *mechanical* properties of the air and all other elastic fluids. The word pneumatics is derived from the Greek ($\pi\nu\epsilon\upsilon\mu\alpha$) which signifies breath or air.

2. *The air is a thin gaseous substance* which rests upon the surface of the earth, enveloping it on every side, and revolving with it on its axis. It extends upwards for a considerable

height and with the clouds and vapours that float in it is called the ATMOSPHERE from ατμος (atmos), vapour, and σφαιρα (sphaira) a sphere.

3. HEIGHT OF THE ATMOSPHERE—The height of the atmosphere has never been exactly ascertained; it is, however observed, that at a greater height than forty-five miles it does not refract the rays of light from the sun, and this height is usually considered to be the height of our atmosphere.

4. THE AIR IS A TRANSPARENT BODY, and consequently *invisible*; but it may be felt by its motion when it blows upon us. By waving our hands we may easily discover its presence, and that it is a fluid body in which we live and move, as fish do in water. The reason why we cannot see the air is because of the minuteness of its particles, which permit the rays of light to pass through them, without reflecting any back to us.

5. THE AIR IS A MATERIAL BODY, and is, therefore, subject to the laws of MATTER, and has the usual qualities appertaining to MATTER, namely, *weight, density, elasticity, compressibility, impenetrability, inertia, mobility*.

6. COLOUR OF THE AIR.—When we look at the sky on a clear day, it appears like a light blue arch set over our heads. But this is not the case; there is no blue dome above us; and when the sky is viewed from any elevated region of the earth, as the top of a high mountain, it appears *dark* or *black*. The apparent blue colour of the air is produced by the refraction of light (see Optics). The air, therefore is without colour.

7. WEIGHT AND PRESSURE OF THE AIR.—The weight of the atmosphere exhibits itself like the weight of other material bodies by *pressure*. Like water it presses equally in all directions, and we are not directly sensible of this pressure; but if the air be removed from any portion of our bodies, its pressure is immediately proved: for instance, if the air be withdrawn from the inside of a thimble placed on the fleshy part of the arm, the *pressure* of the air *on* the thimble not being counterbalanced by air within, causes it to be forced down with great force upon the skin.

8. PHENOMENA RESULTING FROM ATMOSPHERIC PRESSURE.—There are many familiar phenomena which are the immediate consequence of the air being a heavy fluid. Thus, for instance, if we shut the nozzle and valve hole of a pair of

bellows, after having squeezed the air out of them, if they are perfectly air tight we shall find that a very great force (even some hundreds of pounds) is necessary for separating the boards. They are kept together by the pressure of the heavy air, which surrounds them in the same manner as if they were surrounded by water.

9. The common leather sucker with which boys raise stones, acts also from the pressure of the atmosphere. The leather being made pliable by water, is pressed carefully down upon the stone. This pressing of the leather excludes the air from between the leather and the stone; and by pulling the string fastened to the middle of the leather, a vacuum is left underneath its centre; consequently the weight of the air about the edges of the leather not being counterbalanced by any air between it and the stone, enables the boy to lift it.

10. To *atmospheric pressure* is to be referred the well known fact, that a cask will not run by the cock unless a hole be opened in some other part. If the cask be not quite full, indeed, some liquor will be discharged, but it will stop after a time till an opening be made, which lets in the air. For the same reason a small hole is made in the lid of a tea-pot.

11. To the same cause must be ascribed the very strong adhesion of snails, perriwinkles, limpets, and other *univalve* shells to rocks. The animal forms the rim of its shell so as to fit the shape of the rock to which it intends to cling; it then, by means of some muscular exertion, produces a vacuum, and the pressure of the air causes it to adhere firmly. In the same manner do the polypus, the lamprey, and many other animals adhere with that firmness which we so frequently observe. Those animals also which possess the power of walking in opposition to gravity, as flies for example, on the ceiling of a room, or on the smooth perpendicular surface of a looking glass, owe the facility of support to the peculiar construction of their feet, which enables them to produce a vacuum by mere muscular exertion.

12. The pressure of the air may be shown by a simple experiment. Place a card on a wine glass filled with water; then invert the glass, the water will not escape, the pressure of the atmosphere on the outside of the card being sufficient to support the water; or invert a tall glass jar in a dish of

water, and place a lighted taper under it: as the taper consumes the air in the jar, the water from the pressure without *rises up* to supply the place of the air removed by the combustion. In the operation of cupping, the operator holds the flame of a lamp under a bell-shaped glass; the air within this being rarified and expanded, a considerable portion is driven off. In this state the glass is pressed down to the part, and as the inward air cools it contracts, and the glass adheres to the flesh by the difference of the pressures of the internal and external air.

13. AGGREGATE PRESSURE OF THE AIR ON THE GLOBE.—The whole of the atmosphere exerts the same pressure on the surface of the globe, as if, instead of air, it were enveloped with water to the height of thirty-four feet above its surface. or with quicksilver to the height of about thirty inches. That pressure has been calculated to be equal to a weight of 12,022,560,000,000,000,000 pounds (5,367,214,285,714,285 tons), or to be the same as that which would be exerted by a globe of lead sixty miles in diameter.

14. To state this fact in another way, this pressure is equal to a weight of fifteen pounds to every square inch of surface, that is, a column of air an inch square, and reaching from the surface of the earth to the greatest height of the atmosphere, weighs fifteen pounds, and presses as much on the surface which it covers as if there were no air, but a weight of fifteen pounds in its place, and is of the same weight as a column of *mercury* an *inch* square and thirty *inches* high, or a column of *water* an inch square and thirty-four *feet* high. It is therefore a frequent expression that the air can support columns of water or of mercury these respective heights.

15. The air is said to press upon the substances which it surrounds, but it cannot be said, properly speaking, that a thing presses on a substance which by the same thing is equally pressed off. It is only when the counterbalance is destroyed by removing the air from a thing that it can be said to be pressed upon by the air. The pressure of the air on the body of a man is equal to thirteen tons when the barometer stands at 29.5. The number of square inches on the surface of a man's body is 2088, and the pressure upon each inch 15 lbs. This pressure would crush a man but that there is

ATMOSPHERIC PRESSURE.   77

the pressure of the air within, which is equal to it, and this prevents any disturbance of our animal economy.

16. From this weight of the body of the atmosphere, which is equal to that of thirty-three feet of water, depends the construction of the common pump. Galileo discovered that it was impossible to raise water higher than thirty-three feet; he therefore concluded that a column of water thirty-three feet high, was a counterpoise to a column of air reaching to the top of the atmosphere. (See Hydraulics.)

17. It is the same pressure that makes the mercury keep up in the barometer. Galileo's pupil, Torricelli, considered that as mercury was fourteen times as heavy as water, a column of that fluid need only be one-fourteenth of the length of one of water, to form an equal counterpoise to the pressure of the air; and accordingly, having filled with mercury a glass tube about three feet long, hermetically sealed at one end, he inverted it into a small basin of mercury, and found as he expected, that the mercury subsided to the height of about 29½ inches, and there remained suspended, leaving at the top of the tube a space, or perfect vacuum, which has been called the *Torricellian vacuum.** In the instrument thus formed, which

* Dr. Cotes has shown that if altitudes in the air be taken in *arithmetical* proportion, the rarity of the air will be in geometrical proportion. For instance—

At the altitude of	Miles above the surface of the earth the air is—		times thinner and lighter than at the earth's surface.
7		. 4	
14		. 16	
21		. 64	
28		. 256	
35		. 1,024	
42		. 4,096	
49		. 16,384	
56		. 65,536	
63		. 262,144	
70		. 1,048,576	
77		. 4,194,304	
84		. 16,777,216	
91		. 67,108,864	
98		. 268,435,456	
105		. 1,073,741,824	
112		. 4,294,967,296	
119		. 17,179,869,184	
126		. 68,719,476,736	
133		. 274,877,906,944	
140		1,099,511,627,776	

And hence it is easy to prove by calculation, that a cubic inch of such

ATMOSPHERIC PRESSURE.

is a barometer, the mercury will stand in the glass tube to the height of twenty-nine or thirty inches, a little more or less, according to the weight, and consequently the pressure of the atmosphere at the time. The reason of this is, that the pressure of the whole atmosphere will not raise a column of mercury higher than about thirty inches. That is, the pressure of the atmosphere is equal to the pressure of a column of mercury that number of inches in height, in the same way as the pressure of the atmosphere is equal to a column of water thirty-two or thirty-three feet high.

18. In the action of the pump the atmosphere presses equally on the whole surface of the water in the well, until the rod of the pump is moved; but by forcing the rod down, the bucket compresses the air in the lower part of the pump tree, which being elastic, forces its way out of the tree through the valve, so that when the bucket is again raised, that part of the pump tree under the bucket is void of air, and the weight of the atmosphere pressing upon the body of water in the well forces up a column of water to supply its place; the next stroke of the pump rod causes another column of water to rise, and so long as the bucket fits the pump tree close enough to produce a vacuum, a constant stream of water may be drawn from below.

air as we breath would be so much rarified at the altitude of 500 miles, that it would fill a sphere equal in diameter to the orbit of Saturn.

At the height of three miles above the surface of the globe, the level of the sea is the height from which such calculations are made, the air is about half as dense or heavy as at the surface, and the density decreases in the following proportion:

	DENSITIES.
Level of the Sea	1
Height of 3 miles	$\frac{1}{2}$
6	$\frac{1}{4}$
9	$\frac{1}{8}$
12	$\frac{1}{16}$
15	$\frac{1}{32}$

If the air in a vessel filled with air at the level of the sea weighs one ounce, it will weigh half an ounce if taken from the height of three miles, a quarter if from the height of six, and so on in the same proportion.

CHAPTER XVIII.

DENSITY OF THE AIR.—ITS ELASTICITY AND COMPRESSI-
BILITY.—FORCE OF CONDENSATION.—CONDENSER.— AIR
GUN, ETC.

19. DENSITY OF THE AIR.—Air is not always of the same density, nor is it equally dense at different heights of the atmosphere. Its density is proportioned to the force that compresses it. It is more dense near the earth's surface than in the upper regions, from its bearing the weight of the air above. Hence, the heights of mountains are measured by the barometer: the first 103 feet the barometer ascends, the mercury falls one-tenth of an inch, 103 feet of air being equal to one-tenth of an inch of mercury at or near the level of the sea. The pressure of the atmosphere decreases in a geometrical progression, as the height of the place of observation increases in an arithmetical progression.

20. Gay Lussac filled a bottle with air at the height of about three miles when in a balloon, and when he descended, took out the stopper under water. The water rushed in and filled half of the bottle; the air in the bottle being compressed into less bulk by the greater pressure to which it was now exposed. This lightness, or thinness of the air in the upper regions is expressed by the term *rarity*, and we say that the air gradually becomes more *rare* as it is farther from the surface, and more *dense* as it is nearer the ground.

21. The different densities of the air may be illustrated by conceiving twenty or thirty equal packs of wool placed one above another; the lowest will be forced into less space, and its parts brought nearer together, and it will be more dense than the next, and that will be more dense than the third from the bottom, and so on till we come to the uppermost, which sustains no other pressure than that occasioned by the weight of the incumbent air.

22. ELASTICITY OF THE AIR.—The air and other gases are generally termed *elastic fluids*. When air is subjected to pressure, it yields to pressure; its particles are brought nearer

to each other, and it is *contracted*, or occupies a smaller space than before, which is easily proved by squeezing in the hand a blown bladder. The elasticity and compressibility of common air may also be shown by inverting a long ale-glass in a basin of water. In this experiment the air will gradually take up less and less room above the surface of the water within the glass, as it is forced down to greater depth. When the weight of the hand is removed, the glass will jump up with considerable force, owing to the elasticity of the air within it.

23. Every particle of the air has this property, and it exerts its force equally in all directions, and when released from the compressing force assumes a spherical figure in the interstices of the body which contains it. On this account, the air rising from a piece of lump sugar when dissolving in water, takes the form of globules, and for the same reason by blowing air through an iron tube into a piece of melted glass at the end of a tube, as in glass-blowing, it assumes a spherical shape.

Glass-Blowers.

24. How far air has a power to dilate itself when all external pressure is removed, is uncertain, although it is known to possess this power in a high degree. In several experiments made by Mr. Boyle, it dilated into nine times its former space, then into thirty, and then into sixty, and lastly into one

ELASTICITY OF THE AIR.

hundred and fifty times; but it was afterwards dilated into 9,000 times its first space, and utimately into 13,679 times its original bulk, and this without the assistance of heat.

25. It is the property of elasticity, that repelling principle acting between the particles of the atmosphere which preserves it in a gaseous form. Were it not for this repulsion, the air would be liquid at the surface of the earth, and the pressure of the columns of the air above would cause the particles of air to approach. Indeed, it has been calculated, that at a depth of about fifty miles below the surface, air would be as dense and heavy as quicksilver, owing to the very great pressure above overcoming the elasticity of the air below. At the surface of the earth, however, this pressure is not sufficient to overcome the elasticity which keeps the particles asunder, although every square inch *there* has a force of about fifteen pounds pressing on it, and hence the air remains gaseous.*

26. There is a beautiful philosophical toy of which the action depends chiefly upon the *elasticity of the air*. If three or four little wax men, made hollow within, and having each a minute opening at the heel, by which water may pass in and out, be placed in a jar as seen in the figure, and adjusted by the quantity of water admitted into them, so that in specific gravity they differ a little from each other; and if then the hand be pressed on the mouth of the jar, which is covered with a piece of skin or India rubber, the figures will be seen to rise or descend as the pressure is gentle or heavy, rising and falling, or standing still, according to the pressure made. The reason of this is, that pressure on the top of the jar condenses the air between the cover and the water surface, this condensation then presses on the water below and influences it through its whole extent, compressing also the air in the

* The law of the elasticity of the air is, that its spring or resistance to compression increases exactly with its density or the quantity of it collected in a given space. Hence, by finding in any case either the density of the air or the spring or the compressing force, we know all the three.

figures, forcing as much more water into them as to render them heavier than water, and therefore heavy enough to sink.

27. This toy proves many things. The *materiality* of the air by the pressure of the hand on the top being communicated to the water below through the air in the upper part of the jar; the *compressibility* of air by what happens in the figures just before they descend; the *elastic* force of air is shown in expansion when, on the pressure ceasing, the water is again expelled from the figures; the *lightness* of the air in their buoyancy. It shows also that in a fluid the *pressure is in all directions*, because the effects happen in whatever position the jar be held. It shows that *pressure is as the depth*, because less pressure of the hand is required, the further the figures descend in the water, and thus it exemplifies fluid support. A young person therefore, familiar with this toy, has learned the leading truths of *Hydrostatics* and *Pneumatics*.

28. COMPRESSIBILITY OF THE AIR.—The compressibility of the air is a consequence of its elasticity, for whatever is elastic is capable of being forced into a smaller space. Air may be condensed by artificial means into fifty thousand times less space than it usually occupies.

29. The instrument used for the purpose of condensing the air is called a "CONDENSER." It consists of a cylinder similar 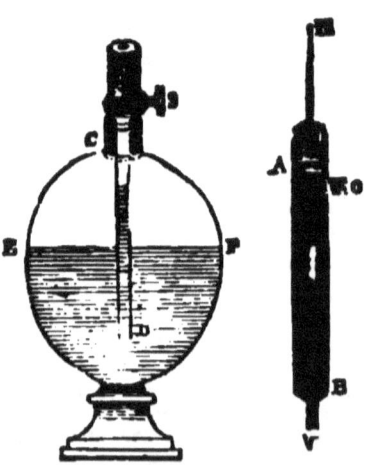 to a common squirt, having an opening in the side *o* to let in the air, and a valve at the end B, which opens downwards. To it is fitted a piston H, which forces the air through the valve B. This being screwed on the air-tight vessel C E F, half filled with water, may be then worked by the hand, and any quantity of air driven into it, which may be seen to rise through the water at each successive stroke. When the piston is drawn up, the valve B, closes, preventing the return of the

air. If, when the condensation be at its height, the piston be taken off, and a small jet tube screwed in its place, and then the cock be turned, the pressure of the air upon the surface of the water will force it up in the form of a jet d'eau, in a height proportionate to the amount of condensation.

30. Atmospheric pressure changes the temperature at various altitudes. If a gallon of air at the surface of the earth contain a certain quantity of heat, this must be diffused equally through the space of a gallon; but if the air be then compressed into one-tenth of the bulk there will be ten times as much heat in that *tenth* of the space it occupies, as there was before, and the increase will effect the thermometer. In like manner, if by taking off pressure, one gallon be made to dilate to 10 gallons, the heat will be in the same proportion diffused, and any one part will be proportionally colder than before. It is well known that air may be so much compressed under the piston of a close syringe that the heat in it similarly concentrated becomes intense enough to inflame tinder attached to the bottom of the piston. This contrivance is occasionally employed to obtain the instantaneous light, and is called the syringe match.

31. THE AIR-GUN.—This instrument owes its power to the expansive force of condensed air. In appearance the air-gun resembles a common musket with the addition of a round copper ball. The ball is hollow; into it the air is forced by means of a syringe, and thus condensed; a portion is then allowed to escape each time the trigger is drawn, so that it presses against the bullet precisely in the same way as common gunpowder.

32. Into the copper ball (*c*) enough air can generally be con-

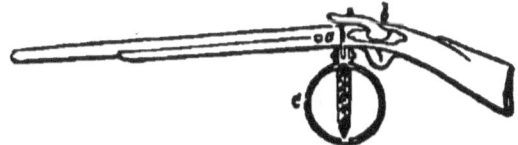

densed to serve for fifteen or twenty separate charges; the force, however, with which they will propel the bullets diminishes each time, because the condensation becomes less upon the loss of every portion of air, so that after a number of discharges the bullet will be projected only to a short distance.

33. The Wind-Gun already alluded to is a more formidable instrument. In this there is a magazine of bullets, as well as another of air, and when it is properly charged, the bullets may be projected one after another as fast as the gun can be cocked and the pan opened. The syringe is fixed on the ball of the gun, by which it is easily charged, and may be kept in that state for a long time.

CHAPTER XIX.

THE AIR-PUMP.—EXPERIMENTS TO PROVE THE MECHANICAL PROPERTIES OF AIR.

34. THE AIR-PUMP.—Most of the important facts connected with the mechanical properties of the air with which we are acquainted, may be referred to experiments with the air-pump, for, prior to its invention, the mechanical properties of the atmosphere were known only by a few isolated phenomena, such as have been already detailed; but the air-pump enables the experimentor to demonstrate to a greater nicety the facts he propounds.

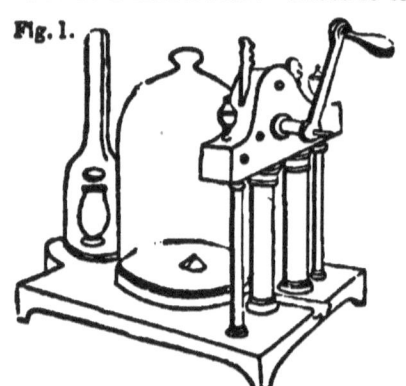
Fig. 1.

35. The air-pump is constructed on similar principles to the water-pump, and whoever understands the one will be at no loss to comprehend the other. It was invented by Otho Gueric, of Magdeburg, but was much improved by Mr. Boyle, to whom we are indebted for the greatest part of our knowledge of the

THE AIR-PUMP.

properties of air, as demonstrated in his wonderful experiments with this interesting machine.

36. Under the receiver of an air-pump (the glass cover shown in the cut) thoroughly exhausted, rare and dense bodies fall with equal swiftness. Most animals die in a minute or two, but some amphibia live hours. Frogs and adders continue alive a long time; vegetation however stops; combustion ceases; gunpowder will not explode; heat is slightly transmitted; a bell sounds faintly; magnets are powerless; glow-worms give no light; and watery and other fluids turn to vapour.

37. There have been several modifications in the construction of the air-pump. Figure 2, is a sectional drawing of the one in most general use. R is a glass *receiver*, having its lower edge ground smooth so as to rest in close contact with the pump plate S S, which must be either rubbed with some unctuous matter or covered with a piece of wet leather, in order to exclude the admission of any external air. In the plate S S is a small aperture A communicating by the tube T T with the two exhausting barrels, B b. C is a stop-cock, by which the communication between the barrels and the receiver may be cut off at pleasure. V V are two air-tight valves opening *upwards*, so that the air can pass from the tube T into the barrels, but cannot return again. In the barrels B b are two pistons P P, so closely fitted that they are perfectly air-tight; each piston contains a valve opening *upwards*, similar to those at V V.

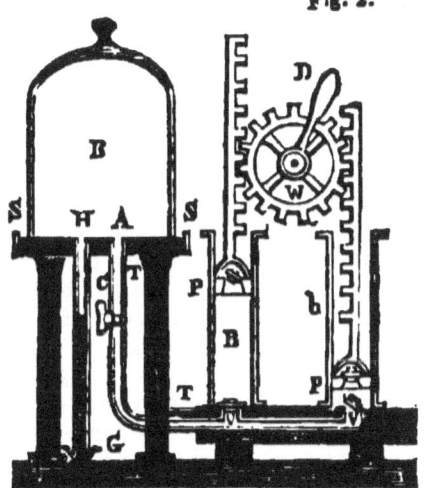

Fig. 2.

The piston rods are furnished with teeth, and are worked by a pinion-wheel W. To exhaust the air from the receiver the stop-cock C must be set open, then by turning the handle D *backwards,* the piston will be depressed, and the air contained in the barrel B will escape through the valve in the piston P; for as the valve V opens upward, it cannot pass downward through it. While the piston P is descending, the other piston P is ascending in the barrel *b*; and as no air can enter the barrel through the valve in the piston, the vacuum formed by the ascent of the piston is supplied by air rushing from the receiver through the tube T and the valve V into the barrel *b*. The elastic spring of the air is the cause of this, which, expanding, fills the vacuum in the barrel *b*. When the handle D is turned *forward,* the piston P will ascend, and the air from the receiver fill the barrel B, and the piston P will descend again and expel the air contained in the barrel, through the valve in the piston P. Thus there is a vacuum made in each barrel, when its piston is raised, and that vacuum is constantly replenished by the elastic air contained in the receiver. The pistons, by the backward and forward motion of the handle, continue to ascend and descend, thereby expelling at every stroke, a portion of the air from the receiver, until at length it becomes so dilated, and its spring so far weakened that it has not sufficient force to raise the valves V V, and then no more can be exhausted.

38. Hence it is obvious, that though the air may be drawn from the receiver sufficiently for all practical purposes, it is impossible to produce a *philosophical vacuum* in it; for the air will but only pass into the barrels until its elastic force is so far reduced that it is unable to raise the valves V V. Thus there must always remain a certain portion of air in the receiver, though so inconsiderable in quantity, that it forms no obstacle to obtaining accurate results from any practical experiments. G H is a *guage* attached to the air-pump by which the exact state of the rarefaction of the air in the receiver is indicated during the process. It consists of a small glass tube open at both ends; the upper end communicates with the receiver at H, the lower end G is immersed in a cistern of mercury, then as the air becomes exhausted in the receiver, its incumbent weight being removed from the column in the tube, and no longer balancing the external pressure of the atmosphere on the cistern G, the former continues gradually to rise as the air in the receiver increases in rarefaction; but as the air can never be wholly expelled from the receiver, the

mercury will never rise in the guage as high as in the barometer tube, because there will be a pressure of that portion of the atmospheric air which remains in the receiver on the column of mercury; therefore the difference in height between the mercury in the guage, and the height of the mercury in the barometer (say 30 inches), will be sufficient distinction to determine the exact degree of rarefaction produced in the receiver.

EXPERIMENTS WITH THE AIR-PUMP.

39. TO PROVE THAT AIR HAS DOWNWARD PRESSURE.—Place the hand on the hole in the middle of the plate, and it will be forced down by the air above, as the air beneath is withdrawn; or exhaust the glass receiver, and it will be held down by the pressure of the air. Remove the receiver, and over the orifice A place a glass tube open at both ends; over one end tie a piece of skin or bladder, and exhaust the air from within, and the pressure of the air will force the bladder down with a loud report.

40. TO PROVE THAT AIR ITSELF HAS WEIGHT.—Take a Florence flask, fitted up with a screw and a fine oiled silk valve. Screw the flask on the plate of the air-pump; exhaust the air, take it off the plate and weigh it. Suppose it to weigh three ounces and five grains. Now let the air into the flask and weigh it again; the weight between the weight now and when tried before, is the weight of the quantity of air contained in the bottle. Suppose it to weigh 3 ozs. $19\frac{1}{4}$ grs; the air will of course weigh $14\frac{1}{4}$ grains. The flask holding a quart, wine measure.

41. To find the weight of air contained in a room which measures in length 40 feet, and in breadth 20 feet, and in height 15 feet; multiply these three numbers, $40 \times 20 = 800 \times 15 = 12,000$. Now, as a cubic foot of water weighs 1,000 ounces, the weight of a room full of water would be $12,000 \times 1000 = 12,000,000$: but air being eight hundred times lighter than water, the air in the room will weigh 12,000,000, divided by 800, which equals 15,000 ozs., or 937 lbs. 8 ozs.

42. TO PROVE THE SPECIFIC GRAVITY OF AIR COMPARED WITH THAT OF WATER.—Exhaust the flask again of its air,

and putting the neck of it under water, lift up the silk valve and fill it with water. Now dry the outside very thoroughly and weigh it, and it will weigh twenty-seven ounces. Subtract the weight of the flask, and reduce the remainder into grains, and divide by 14½; and the specific gravity of water compared with that of air will be obtained, that is, it will be found that water is something more than eight hundred times heavier than air. Hence then the specific gravity of water is always put at 1, that of air must be as $\frac{1}{800}$th at least. But following more accurate experiments, the specific gravity of air is 800 times less than that of water, only when the barometer stands as high as thirty inches.

43. TO PROVE THAT AIR IS ELASTIC.—Place a bladder, out of which all the air has apparently been squeezed, under the receiver; upon it lay a weight. Exhaust the air, and it will be seen that the small quantity of air left within the bladder will so expand itself as to lift the weight. Put a corked bottle under the receiver and exhaust the air, and the cork will fly out.

44. TO PROVE THE RESISTANCE OF THE AIR.—To shew the resistance of the air a double mill is constructed, as in the cut. It consists of two sets of vanes, one made to strike the air with the broad, and the other with the narrow part. The mills, if set going in the air, will not continue the same length of time in motion; for the one d will drive the air with the broad part of the vane and feel its resistance. The vanes a will cut the air with their edges, and continue to turn after the other has stopped; place them under the receiver and exhaust the air. If they are then put in motion, they will continue turning the same length of time.

45. If a guinea and a feather be dropped from the hand at the same time in the open air, the guinea comes to the ground first, because the feather presents a larger surface to the air in proportion to its weight; but under the exhausted receiver they will both reach its bottom at the same moment. (See fig. A.)

46. Place a nicely adjusted balance in the receiver of an air-pump; suspend at one end of the receiver a small piece of lead,

and on the other a piece of cork that will counterpoise it; exhaust the air, and when a vacuum is formed in the receiver, the cork will preponderate and the lead ascend, thereby making cork appear heavier than lead. (See fig. B.)

47. This is accounted for by considering the resistance of air, for when the two substances were weighed in common air, the cork occupying the greater space met more resistance in its descent than the comparatively small body of lead; but being weighed in a vacuum, the resistance of the air is removed, and both bodies descending by the laws of gravity alone, the cork will more than counterbalance the lead. *This shews that air resists bodies in motion, and that bodies equal in weight meet with different degrees of opposition, according as they present greater or less surface to the air.*

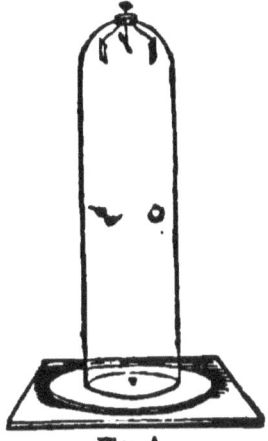

Fig. A.

48. The annexed figure is another beautiful illustration of the resistance of the air. Procure a tall glass vessel, A, closely filled to the plate B. Into the plate B fix a tube C E, which must be furnished with the stop-cock D; screw the end E of the tube into the exhausting tube of the air-pump, then expel the air from the receiver A, and when the exhaustion is completed, close the stop cock D, unscrew the tube E from the air-pump, immerse its lower end into the basin of water F, open the stop cock, and the pressure of the air on the water in the basin will force it upwards through the tube into the vessel A; and there being no air to resist its ascent, it will

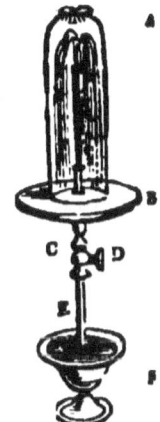

Fig. B.

form itself into a beautiful jet, and continue to play till the equilibrium is restored.

49. To PROVE THE PRESSURE OF THE AIR.—Two hollow hemispheres constructed of brass, are so formed that when placed mouth to mouth they shall be in air-tight contact. They are furnished with handles, C D, one of which may be unscrewed. In the neck to which this handle is screwed is a tube furnished with a stop cock. The handle being unscrewed, let the hemisphere be screwed on the pump-plate, and the other hemisphere being placed over it, let the stop cock be opened so as to leave a free communication between the interior of the sphere and the exhausting tube of the air-pump; the air is to be now exhausted from the interior of the cups. When this is done the stop-cock is again turned so as to prevent the air getting in; the cup is then taken off the plate. The pressure of the air is now so great all round the sphere or cup as to prevent its two hemispheres from being separated. They are pressed together by a force amounting to fifteen pounds for every square inch of the section. If the diameter of the sphere be six inches, its section through the centre will be about 28 square inches. If 28 be multiplied by 15, we shall obtain 420, which is the amount of the force with which the hemispheres will be held together, and if one of the handles be placed on a strong hook, and a weight of 400 pounds be suspended from the other, the weight will be supported by the pressure of the atmosphere.

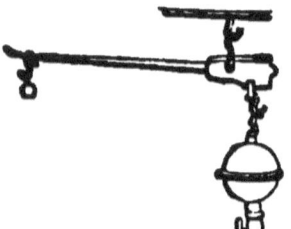

50. Take a receiver, having a brass cap fitted to the top with a hole in it; fit one end of a dry hazel branch about an inch long, tight into the hole, and the other end tight into a hole quite through the bottom of a small wooden cup; then pour some quicksilver into the cup, and exhaust the receiver of air, and the pressure of the outward air on the surface of the quicksilver, will force it through the pores of the hazel, from

whence it will descend in a beautiful shower, into a glass cup placed under the receiver to catch it.

51. Put a wire through the collar of leather on the top of the receiver, and fix a bit of dry wood on the end of the wire within the receiver; then exhaust the air, and push the wire down, so as to immerse the wood into a jar of quicksilver on the pump-plate; this done, let in the air, and upon taking the wood out of the jar and splitting it, its pores will be found full of quicksilver, which the force of the air, upon being let into the quicksilver, drove into the wood.

52. Set a square phial upon the pump-plate, and having covered it with a wire cage, put a close receiver over it, and exhaust the air from the receiver; in doing which, the air will also make its way out of the phial, through a small valve in its neck. When the air is exhausted, turn the cock below the plate to re-admit the air into the receiver; and as it cannot get into the phial again, because of the valve, the phial will be broken into some thousands of pieces by the pressure of the air upon it. Had the phial been of a round form, it would have sustained this pressure like an arch, without breaking; but as its sides are flat it cannot.

53. If a rat, mouse, or bird, be put under a receiver, and the air be exhausted, the animal will be at first oppressed as with a great weight, then grow convulsed, and at last expire in all the agonies of a most bitter and cruel death. But as this experiment is exceedingly wicked, it is common to substitute a machine called the *lungs-glass*, in place of the animal.

54. If a butterfly be suspended in a receiver, by a fine thread tied to one of its horns, it will fly about in the receiver as long as it continues full of air; but if the air be exhausted, though the animal will not die, and will continue to flutter its wings, it cannot remove itself from the place where it hangs, in the middle of the receiver, until the air be let in again, and then the animal will fly about as before.

55. Put a shrivelled apple under a close receiver, and exhaust the air; then the spring of the air within the apple will plump it out, so as to cause all the wrinkles to disappear; but upon letting the air into the receiver again, to press upon the apple, it will instantly return to its former decayed and shrivelled state.

56. Take a fresh  egg, and cut off a little of the shell and film from its smaller end, then put the egg under a receiver, and pump out the the air; upon which all the contents of the egg will be forced out into the receiver, by the expansion of a small bubble of air contained in the great end, between the shell and film.

57. Put some warm beer into a glass, and having set it on the pump, cover it with a close receiver, and then exhaust the air. Whilst this is doing, and thereby the pressure more and more taken off from the beer in the glass, the air therein will expand itself, and rise up in innumerable bubbles to the surface of the beer; and thence it will be taken away with the other air in the receiver. When the receiver is nearly exhausted, the air in the beer, which could not disentangle itself quick enough to get off with the rest, will now expand itself so as to cause the beer to have all the appearance of boiling, and the greatest part of it will go over the glass.

58. Put some water into a glass, and sink an apple to the 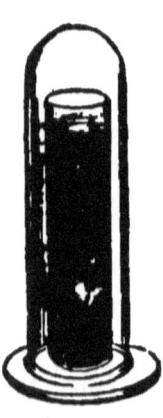 bottom of the water; then cover the glass with a close receiver, and exhaust the air; upon which the air in the apple, having liberty to expand itself, will come out plentifully in small bubbles. The same effect will take place if a small piece of dry wainscoat or other wood be placed at the bottom of the water; a cubic inch of dry wainscoat will continue bubbling for near half an hour together.

59. Set a lighted candle upon the pump, and cover it with a tall receiver. If the receiver holds a gallon of air, the candle will burn a minute; and then, after having gradually decayed from the first instant, it will go out; which shows that a constant supply of fresh air is as necessary to feed flame as to support animal life.

60. The moment when the candle goes out, the smoke will be seen to ascend to the top of the receiver, and there it will form a sort of cloud; but upon exhausting the air, the smoke will fall down to the bottom of the receiver, and leave it as clear at the top as it was before it was set upon the pump. This shows that smoke does not ascend on account of its being positively light, but because it is lighter than air; and its falling to the bottom when the air is taken away, shows that it is not destitute of weight. So most sorts of wood ascend or swim in water; and yet there are none who doubt of the wood's having gravity or weight.

61. Set a receiver which is open at top, on the air-pump, and cover it with a brass plate and wet leather; and having exhausted it of air, let the air in again at top through an iron pipe, making it pass through a charcoal flame at the end of the pipe; and when the receiver is full of that air, lift up the cover. If a candle be let down into that air, it will go out directly; but by letting it down gently, it will drive out the impure air, and good air will take its place.

62. Set a bell on the pump-plate, having a contrivance so as to ring it at pleasure, and cover it with a receiver; then make the clapper strike against the bell, and the sound will be very well heard; but, exhaust the receiver of air, and then the clapper, although seen to strike against the bell, will scarcely be heard; which shows that air is absolutely necessary for the propagation of sound.

63. If a glass vessel containing water in which a couple of fish are put, be placed under the receiver; upon exhausting the air the fish will be found unable to keep at the bottom of the glass, owing to the expansion of the air

in the the air bladders of the fish, and they will, consequently, rise and float, belly upwards, on the surface of the water.

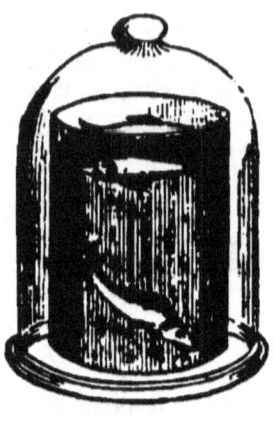

CHAPTER XX.

OF THE BAROMETER, THERMOMETER, HYGROMETER, AND HYDROMETER.

64. THE BAROMETER, as we have seen at page 77, is a column of mercury supported in a tube by the pressure of the atmosphere, and therefore indicating most exactly that degree of pressure. It consists of a glass tube, about two-tenths of an inch in diameter in the bore, and at least thirty-four inches long. It is hermetically sealed at one end, and open at the other. In forming the barometer the maker first fills the tube with quicksilver, and then putting his fingers in the open end so as to prevent any from running out, he inverts the tube into a small cup of quicksilver and withdraws his finger. The mercury now subsides three or four inches, above the top of which in the tube is a perfect vacuum, and

when this tube is fixed to a graduated frame, we have a barometer. The mercury in the glass tube will now stand at the height of twenty-nine or thirty inches, a little more or a little less, according to the state of the air. The reason of this is, that the pressure of the whole atmosphere will not raise a column of quicksilver higher than about thirty inches; that is, the pressure of the atmosphere is equal to the pressure of a column that number of inches in height, in the same way as the pressure of the atmosphere is equal to a column of water thirty-two or thirty-three feet high. (See *Section* 16, 17.)

65. STANDARD ALTITUDE.—The height of mercury in the tube is called the *standard altitude*, which in this country fluctuates between twenty-eight and thirty-one inches, and the difference between the least and the greatest variation is called the SCALE OF VARIATION. At the equator and near the tropics there is little or no variation in the height of the mercury in the barometer in all weathers. This is the case at St. Helena. At Jamaica the variation very rarely exceeds three-tenths of an inch. At Naples about one inch. Whereas in London it is nearly three inches, and at St. Petersburgh it is three inches and a half.

66. AIR PUMP BAROMETER.—A barometer connected with the air-pump indicates exactly the progress and degrees of exhaustion in the receivers. When the mercury falls to half its height it shows that half of the air is extracted, and so for all other proportions. The barometer usually connected with the air-pump is in the form of a bent tube, and is called the syphon guage.

67. WHEEL BAROMETER.—Besides the common barometer already mentioned, there is one called the wheel barometer, which is employed in the weather glasses which have an index in the form of a clock face. It consists of a bent tube filled with mercury, which rises to within a small distance of the top where there is a vacuum. A slender thread having a small weight attached to each end, as seen in the drawing, passes over a pulley, which carries the index pointer. Now, when the weight of the air increases, it presses with greater force upon the surface of the mercury at D, fig. 1, which consequently sinks at D and rises at A. The small weight E

WHEEL BAROMETER.

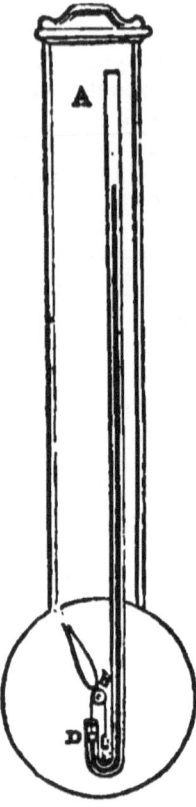

Fig. 1.

being a little heavier than D, follows the mercury. The end of the string B then descends, the index moves over the graduated scale, and thus indicates the degree of pressure which has taken place, and shows to the observer the prognostics of fair, set fair, rain, stormy, &c. as at fig. 2.

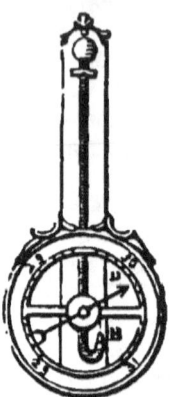

Fig. 2.

RULES FOR PREDICTING THE STATE OF THE WEATHER.

68. In very hot weather, the falling of the mercury indicates thunder.

69. In winter, the rising presages frost; and in frosty weather, if the mercury falls three or four divisions, there will be a thaw. But in a continued frost, if the mercury rises, it will certainly snow.

70. When wet weather happens soon after the depression of the mercury, expect but little of it; on the contrary, expect but little fair weather, when it proves fair shortly after the mercury has risen.

71. In wet weather, when the mercury rises much and high, and so continues for two or three days before the bad weather is entirely over, then a continuance of fair weather may be expected.

72. In fair weather, when the mercury falls much and low, and thus continues for two or three days before the rain comes, then a deal of wet may be expected, and probably high winds.

73. The unsettled motion of the mercury denotes unsettled weather.

74. The words engraved on the scale are not so much to be attended to, as the rising and falling of the mercury; for if it stand at *much rain*, and then rises to changeable, it denotes

fair weather, though not to continue so long as if the mercury had risen higher. If the mercury stands at fair, and falls to changeable, bad weather may be expected.

75. In winter, spring, and autumn, the sudden falling of the mercury, and that for a large space, denotes high winds and storms: but in summer it presages heavy showers, and often thunder. It always sinks lowest of all for great winds, though not accompanied with rain; but it falls more for wind and rain together, than for either of them alone.

76. If, after rain, the wind change into any part of the north, with a clear and dry sky, and the mercury rise, it is a certain sign of fair weather.

77. After very great storms of wind, when the mercury has been low, it commonly rises again very fast. In settled fair weather, except the barometer sink much, expect but little rain. In a wet season, the smallest depression must be attended to; for when the air is much inclined to showers, a little sinking in the barometer denotes more rain. And in such a season, if it rise suddenly fast and high, fair weather cannot be expected to last more than a day or two.

78. The greatest heights of the mercury are found upon easterly and north-easterly winds; and it may often rain or snow, the wind being in these points while the barometer is in a rising state, the effects of the wind counteracting; it is therefore not a *sure* weather guide.

79. MEASURING OF HEIGHTS.—The barometer has been applied with success for the purpose of measuring the altitudes of mountains. It is ascertained that in ascending lofty eminences, the mercury in the barometer will fall about one-tenth of an inch for every 103 feet of perpendicular height. Thus, if a person ascends a high hill, and taking a barometer with him, finds that the mercury has fallen 1½ inch, he may conclude that the hill is about 1,500 feet high. See *Section* 19.)

80. THE THERMOMETER.—The thermometer is also a glass tube, with a round bulb at one end, the other being closed by the joining of the glass, and the whole is thus, technically speaking, *hermetically* sealed. The bulb contains mercury, and the top of the tube is a vacuum. If this tube be placed in powdered ice, as the mercury contracts with cold it will

THE HYGROMETER.

sink low in the tube; at this point a mark is made to represent the freezing point, and this in Fahrenheit's thermometer, the one used in England, is marked 32 degrees. If the thermometer be now immersed in boiling water, the mercury will rise to a certain height in the tube, this is called the boiling point, and is marked 212 degrees. The different degrees between these are usually designated 55 temperate, 76 summer heat, 98 blood heat, 112 fever heat, 176 spirits boil. At the bottom of the scale an 0 is put for zero, or extreme cold.

81. There are three Thermometrical scales in general use by which mercury determines heat, from its freezing point at 40° below zero, to 600°, where it boils. They are *Fahrenheit's*, *Reaumer's*, and the *Centigrade*.

82. To convert degrees of *Fahrenheit's* thermometer into *Reaumer's*, subtract 32, multiply by 4, and divide by 9. To convert degrees of *Reaumer* into *Fahrenheit*, multiply by 9, divide by 4, and add 32.

83. In the Centigrade much used, 0 or zero of Fahrenheit, is—17.78, and 32 of Fahrenheit is 0 of centigrade; 68° of Fahrenheit is 20° of centigrade; 104° of Fahrenheit is 40 of centigrade; every 18° of Fahrenheit being 10° of centigrade throughout.

84. Water *freezes* at 32° of Fahrenheit, 0 of Reaumer, and 0 of centigrade; and it boils at 212° of F., 80° of R., and at 100° of centigrade; so that the distance is 180° of F., 80° of R., and 100° of centigrade.

85. THE HYGROMETER is an instrument contrived for measuring the different degrees of moisture in the atmosphere. The well known contrivance called by some the weather house, is an instrument of this kind. When the air is very moist it therefore *denotes*, not *foretels*, wet weather, then the man comes out; it must be remembered that when the barometer falls the air is *dryest and lightest*; and in fair weather when the atmosphere is dry the woman makes her appearance. The two figures in this toy are placed on a kind of lever, which is sustained by catgut, and catgut is very sensible to damp, twist-

ing and shortening by moisture, and untwisting and lengthening as it becomes dry.

86. A simple hygrometer may be made by a piece of catgut and a straw. The catgut twisted is placed through a hole in a dial like the dial of a watch. In this is stuck a small straw. In dry weather the catgut curls itself up, in wet weather it relaxes, and the straw is consequently turned to one side or the other.

87. Another hygrometer may be made by stretching a piece of whipcord or catgut over five pulleys, B C D E, as seen in the engraving. To its lower end is placed a little weight W sliding up and down by the side of a graduated scale, and as according to the moisture or dryness of the air the string shortens or lengthens, of course the weight rises higher or lower.

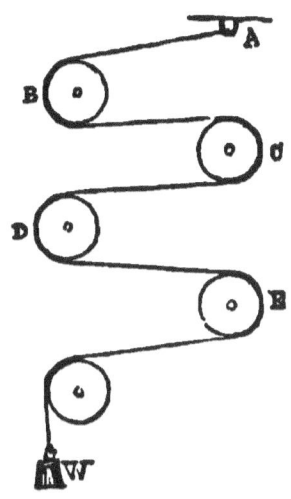

88. A hygrometer may also be made by a piece of sponge A being nicely balanced at one end on the beam of a balance B B by a weight at the other W. To make the sponge imbibe more moisture than it otherwise would, it should be dipped after being made perfectly clean, into a solution of salt in water. The scale is above at D D, and at the fulcrum is an index finger, which will move backwards and forwards in the scale as the sponge is heavier or lighter by the presence or absence of moisture.

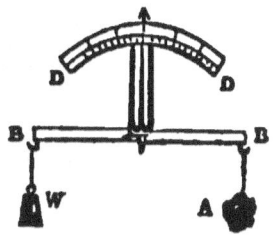

The Diving Bell.

CHAPTER XXI.
AERONAUTICS.

1. Æronautics is the art of sailing in, or navigating a body through the air. In remote ages, the idea of rising in the atmosphere was entertained, and Icarus is said to have risen so high in the air that the sun melted his wings, and he fell into the sea and was drowned. In modern times, when gases lighter than air were discovered, it led to the invention of the balloon.

2. The air balloon is a bag of silk of large dimensions, usually cut in gores, and when extended by its gas, of a pear shape. It ascends in the atmosphere because its whole bulk is much lighter than the air would be in the space it occupies. It is in fact a vessel filled with a fluid which will float in another fluid heavier than itself.

3. When air is made hot it expands itself. This being the case, if 5,000 grains of air fill a space equal to ten feet when the thermometer is at 55° in the open air, 2.500 grains will fill the same space if the air is heated to 200 degrees, and then a weight nearly equal to the air expelled may be carried up with this portion of air. This weight may consist of the balloon itself, its car, and one or more persons.

4. Fire balloons are those raised by heated air, and were the first invented. The first balloon of Stephen and Joseph Montgolfier was a silk bag containing forty feet, which burning paper raised seventy feet. Their next was a bag of six hundred and fifty feet which rose six hundred feet. Their third was thirty-five feet in diameter, and was capable of raising five hundred pounds. It ascended before the public, June 5, 1783. On the 21st of November, Pilatre de Rosier ascended at Paris, and afterwards others with air rarified in the car by heat.

5. Balloons have since been made with silk prepared with liquified india rubber. The car is usually of wicker work, and the balloon has valves for letting out the gas when the æronaut wishes to descend. The gas used for the inflation is often hydrogen gas made in casks by oil of vitrol, water, and zinc. But the gas generally used is the carburetted hydrogen, commonly called coal gas, burnt in the streets, which is obtained from the gas manufactories.

6. The superficies of a balloon are computed by multiplying the square of its diameter by 3.1416, or the cubic contents is the cube, by 0.5236. Taking atmospheric air 1.2 oz. to the cubic foot, we have the weight in air, and then as carburetted hydrogen gas weighs 0.2 oz., the weight of air multiplied by 0.2 gives the power of ascension.

7. In August, 1807, Garnerin passed 7¼ hours in the night in the air, and at 18,000 feet of elevation, in an illuminated balloon, without any personal inconvenience or unusual

phenomena. He saw the sun rise, and in darkness saw little meteors near him. In a second night ascent in a gale of wind he was carried above two hundred and sixty miles in 7½ hours. In another night excursion Madam Blanchard was twenty hours in the air, and in 1838 Mr. Green and two friends passed over from England to France, and after travelling for upwards of 500 miles, descended safely in Germany.

VELOCITY OF BALLOONS.

8. The extraordinary velocity of balloons is to be ascribed to the greater force of uninterrupted air at great elevations, and to the velocity of diagonal ascent. The ordinary rate is from twenty-five to thirty-five miles an hour, but Sadler travelled seventy-four, and Green, on one occasion, no less than ninety-eight. Green states, after 140 voyages, that from 5,000 to 14,000 feet, there is over England a constant N.W. current of six miles an hour, while S.W. sub-currents move from thirty to eighty.

9. Many attempts have been made to improve the shape of balloons, and to render them navigable. A Frenchman proposed to navigate his vessel by four balls exhausted of air, as seen in the engraving. He argued that the diminished weight of the balls would buoy up not only themselves, but the æronaut and his vessel. But it is evident that before balls capable of withstanding the external pressure of the air could be constructed, the materials employed being necessarily of a strong nature, they would turn out to be bulk for bulk heavier than the air. Thus the scheme was abortive, as have been several others for the propulsion or guidance of balloons by means of paddles, wings, or rudders.

THE DIVING BELL.

10. To illustrate the principle of this machine, take a glass tumbler and plunge it into water, with the mouth downwards, and it will be found that the water will not rise much more than half way in the tumbler. This may be made very evident if a piece of cork be suffered to float inside the glass, on the surface of the water. The air within the tumbler does not entirely exclude the water, because air is elastic, and consequently compressible; and hence the air in the tumbler is what is called condensed.

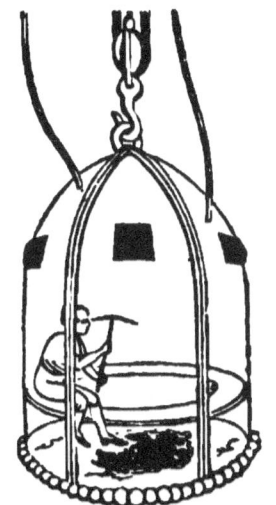

11. A diving-bell is formed on the above principle, but instead of being a glass it is a large wooden or metal vessel sufficiently capacious to hold one or more persons, as may be readily understood by referring to the cut on this page and that at page 100. Air is pumped in from above by means of a forcing air-pump, or condenser, and the persons within are able to breathe freely. Mr. Smeaton's diving-bell, made in 1788, was a square chest of cast iron four and a half feet in height, four and a half feet in length, and three feet wide, and afforded room for two men to work in it. It was used with great success in laying the foundation of the Eddystone lighthouse.

12. Other contrivances have been used for diving-bells within the last thirty years; they have been frequently employed in laying the foundations of buildings under water; and one on an improved principle was constructed in 1812, by the late Mr. Rennie, and employed in Ramsgate Harbour, where it answered so well that the masonry was laid with the utmost precision, and the expense of coffer dams was avoided.

13. The pump connected with the diving-bell requires par-

ticular description. It is composed of a short cylinder of large diameter, and of a piston moved by a lever; the piston has six circular openings shut up in the body of the pump by leather valves, set in copper, which open to permit the entrance of the air when the piston is raised, and which shut when it is depressed. In order to prevent the reflux of the air introduced into the bell, there is a valve at the end of the tube which closes upon the piston when it is raised and opens when it is forced down; by this mode of renewing the air the workmen can remain under water for hours together, without any inconvenience.

14. The bell represented in the cut is of cast-iron, of a single piece. Its length at the base is eight feet, and its breadth five feet. In the upper part are several square holes, in which glass illuminators are fixed, made perfectly air-tight. In the interior of the bell is a seat passing entirely round it, on which the workmen sit. Between the illuminators strong bands of iron are placed, to which is fixed the chain for lowering and raising the machine, and on each side is a leather tube hose, with a copper rim screwed into the bell, through which fresh air is admitted when it is below the surface of the water.

15. Beside the diving-bell, divers are now employed to descend to submarine depths. They are clothed in a water-tight and air-tight dress, having a capacious head-piece, from which flexible tubes ascend to the surface, from which air is supplied by means of a forcing pump. By the means of the flexible tubes, the diver is enabled to walk about beneath the water, having the free use of his limbs, and thus performs the labour of clearing away or of removing masses of wreck, or of recovering long-buried treasure. In Colonel Pasley's late operations on the Royal George, sunk at Spithead in 1782, the greater part of the wreck has been recovered, after blasting by gunpowder, by means of divers.

METEOROLOGY.

Snow flakes magnified.

CHAPTER XXII.

OF EVAPORATION, RAIN, SNOW, HAIL, DEW, FOGS, AND MISTS.

1. Connected in degree with Pneumatics is the science of METEOROLOGY, which treats of the atmosphere, the alterations that take place in it, of its currents of winds, of the variations of its pressure, of the state of the electricity which it exhibits, and lastly, as to the visible phenomena dependent upon these changes.

2. Whatever is engendered in the air which surrounds us, and which appears to be below the moon is a meteor. Meteors are composed of vapours and exhalations.

3. The air, from its dissolving mechanically and chemically various subtances with which it comes in contact, partakes of the nature of these substances. Hydrogen gas is constantly

generated at the earth's surface, and being lighter than common air floats above it, and there produces many combustible phenomena. This chemical and mechanical action constantly going on produce accumulation of the electrical power, the equilibrium of which is restored by the phenomena of thunder and lightning.

4. EVAPORATION.—Water is continually converted into vapour by the sun's action; this vapour is lighter than the atmosphere, and thus rises above the earth's surface. In the colder regions it becomes partially condensed, and then forms clouds. A large portion of the earth's surface being covered with water, it is constantly evaporating and mixing with the atmosphere in this state of vapour.

5. RAIN.—When the heat or density of the air is suddenly disturbed by electricity or other causes, the particles of vapour form themselves into drops too heavy for floating in the atmosphere; they then fall in the shape of drops of water, and are called rain.

6. SNOW and HAIL are the same drops frozen in their descent towards the earth. In the former the frost catches the resolving cloud just as its particles are about to unite into drops, and in the latter the drops are actually frozen in their descent. In the former a regular crystallization takes place of the most beautiful kind, as in the cut at the head of the chapter, and of various geometrical forms.

7. DEW is produced from a quantity of particles of water extremely subtle, that float about in a calm and serene air in the form of vapours, which, being condensed by the coldness of the night, lose by degrees their agitation and unity together, and fall in the morning in small invisible particles, like an extremely fine and delicate rain.

8. The economy of nature in evaporation is very beautiful. The vapours rising from the seas are wafted by the winds in clouds over the land; here by their electricity and heat being disturbed by the warmer surface of the earth and its electric power, they are formed into refreshing rains, water the earth as a gardener would his garden, support vegetation, and falling from little streams to rivers run again to the sea.

9. FOGS are clouds floating on the earth's surface, and clouds are fogs in the higher regions.

10. Mists are collections of vapours which chiefly arise from fenny moist places, and become more visible as the light of day increases.

OF WINDS, MONSOONS, HURRICANES, TORNADOES, AND WATERSPOUTS.

11. When the air is rarified it naturally ascends to the higher regions, and the circumjacent air, which is thicker and heavier, immediately rushes in to supply its place and fill up the vacancy. This motion of the air we call WIND.

12. The rarefaction of the air is caused by the heat of the sun. The greatest heat of the sun being at the equator, cold air will rush towards it from each of the poles. This air will not have the motion of the great body of heated air which is rushing onward with the sun, but it will seem to have a backward motion, and will appear as a wind from east to west.

13. Thus are formed the two great currents of the atmosphere. The one in the torrid zone is from east to west, the other from the poles to the equator. From the combined action of these two are formed the *Trade Winds*.

14. The great trade winds of the Atlantic and Pacific Oceans are termed *constant* winds, because they blow always from the same point; they are *easterly*. At different seasons of the year, however, they vary a little, being rather more N.E. in the northern hemisphere, and rather more S.E. in the southern, when the sun is respectively in those hemispheres.

15. The sun's heat is more strongly reflected when it shines on the land than when it shines on the water. This fact determines the course of the *Monsoons*,—the *periodical* winds of the Indian Ocean. The sun, shining vertically over the great mass of land in Southern Asia, the air over that part becomes more heated than the air over the Indian Ocean. The cold air of the sea then rushes northward, and as this air, coming from the equator, possesses the eastward diurnal motion of the earth, the *Monsoon* so produced is a south-west wind. In the winter, the monsoon is from the north-east.

16. On most of the coasts situated between the tropics the wind blows towards the shore in the day time, and towards the sea by night. These are called the sea and land breezes. The

heat of the sun during the day rarefies the air all over the land, and this causes land breezes. In the night the contrary is the case.

17. If we put in the middle of a large dish of cold water a water plate filled with hot water, the former will represent the ocean, and the other the heated land rarifying the air over it. If a lighted candle be now held over the cold water and blown out, the smoke will then move towards the plate. We may reverse the experiment by filling the outer vessel with warm water, and the plate with cold; the smoke will then move from the plate to the dish.

18. In England, the north east wind generally blows with us during the spring months, and the same wind prevails during the same period in the Northern Pacific. Hence, at that season, the cold air from the north of Europe and America flows into the Atlantic and Pacific, and is the reason of the air's uncommon dryness and density. The variableness of the wind in general is proverbial, and depends probably on a variety of causes, principally on the electrical currents abounding in the air.

19. The force of the wind increases as the square of the velocity, thus, if on a piece of board exposed to a given wind, there be a pressure equal to a pound, and the same board be exposed to another wind of double the velocity, the pressure will be in this case four times greater than it was before.

20. It has been proved that the force of a wind which travels at the rate of forty-five miles an hour, is equal to the perpendicular force of ten pounds avoirdupoise weight on every square foot. Now, if the surface were a large tree, with all its branches and leaves presented to the wind, no one will be surprised that in great storms some of them should be torn up by the roots.

21. The breeze is estimated to press with a force of about two ounces; the fresh gale from eight to sixteen ounces; the strong gale from four to eight pounds; the hurricane from forty-five to ninety pounds on a square foot.

22. HURRICANES are sudden and violent gusts of wind, which come at very irregular periods, and generally continue for a short time.

23. TORNADOES are violent winds attended with particu ar

phenomena, such as draughts, heavy rains, hail, rain, snow, and thunder.

24. WHIRLWINDS.—A whirlwind is formed when gusts of wind come from different quarters at the same time, and meeting in a certain place, the air acquires a velocity or screw-like motion round an axis, which is sometimes stationary, and at others moving in its particular directions.

25. WATERSPOUTS.—The waterspout commonly begins with a small cloud, which the mariners call the squall; this augments in a little time into an enormous cloud of a cylindrical form, or that of a reversed cone, and produces a noise like an agitated sea, sometimes emitting thunder and lightning, and pouring out large quantities of rain or hail, sufficient to overwhelm large vessels, overthrow trees and houses, and everything which stands in the way of its course. Both this and the whirlwind are results arising from the same general cause, and are explicable upon principles furnished by electrical experiments and discoveries.

26. CLASSIFICATION OF CLOUDS.—There are seven classes of clouds :—1, *Cirrus*, or curl cloud, so called from its resemblance to a lock of hair or feather; 2, *Cumulus*, a cloud in round conical masses; 3, *Stratus*, a level sheet; 4, *Cirro-cumulus*, a system of small round clouds; 5, *Cirro-stratus*, a concave or undulating stratus; 6, *Cumulo-stratus*, the cumulus and cirro-stratus mixed; 7, a cumulus spreading out in cirrus, and raining beneath, called NIMBUS.

27. When rain takes place the nimbus cloud is formed; a cumulus becomes at rest, and a cirrus or cirro-stratus settles on it, and the whole change to a cumulo-stratus, first black and then grey. The fall of rain forms cirrus fibres, which afterwards float on small cumuli, and the nimbus rises and is separated.

28. The cirrus, or curling cloud, is always uppermost and often five or six miles high, it portends rain and wind. The cumulus, or *streaked* cloud, is low and massive. The cirro-stratus is long and flat, or in wavy bars, and often in broken patches like a mackerel's back. Clouds are often of an enormous size, 10 miles each way and 2 thick; they are composed of little globules of water at small distances from each other.

ACOUSTICS.

CHAPTER XXIII.

DEFINITION OF SOUND—ILLUSTRATION—SOUND AND NOISE—DEPENDS ON VIBRATION—AN ECHO OR REFLECTION—SOUNDS IN LIQUIDS AND IN SOLIDS—HOW SOUND IS TRANSMITTED—THE STRUCTURE OF THE EAR.

1. The term ACOUSTICS is derived from the Greek word, *akouo*, which signifies *I hear*. It is therefore the science of sound and hearing, and treats of the laws of the production and propagation of sound.

2. SOUND IS HEARD when any sudden shock or impulse is given to the air, or to any other body which is in contact, directly or indirectly with the ear, such as the crack of a whip, the blow of a hammer, the thunder-clap, &c.

3. These impulses, if quickly repeated, cannot be individually attended to by the ear, and hence they appear as one continued sound, of which the pitch or tone depends on the number occurring in a given time; and all continued sound is but a repetition of impulses.

4. If a wheel with teeth be made to turn and to strike any elastic plate, as a piece of quill, with each tooth, it will, when moved slowly, allow the motion of every tooth to be seen and every blow to be separately heard; but with an increased velocity the eye will lose sight of the individual teeth, and the ear ceasing to perceive the separate blows, will at last hear only a smooth continued sound called a tone, of which the character will change with the velocity of the wheel.

5. When a continued sound is produced by impulses which do not, like those of an elastic body, follow in regular succession, the effect ceases to be a clear uniform sound or tone, and is called a noise; such as the sound of a saw or grindstone, the rattling of wheels, the clank of hammers, or the mixed voices of a talking multitude.

6. VIBRATION.—Sound arises from vibrations of the air, as may be seen by the *agitations* of a bell when it is struck, which are communicated to the air in contact with it, or by the vibrations of the water of a musical glass, and by the motions of light bodies laid on strings in concord. In a harp string vibrations may be detected by the eye, and in most other instruments by the touch. These vibrations have been made the subject of calculation, and the lowest note which is perceptible to the human ear has about thirty beats in a second,—the highest thirty thousand, and there is included between these a range of nearly ten octaves.

7. How PROPAGATED.—The shock which causes the sensation of sound, spreads, or is propagated in all bodies, solid or fluid, somewhat as a wave spreads in water, with decreasing strength as the distance increases, but with a velocity nearly uniform, and which in air is about 1142 feet per second.

8. SOUND AN ECHO.—All sound appears to be echo, or reflection, and if not a distant echo, it is only from want of distance. In a real echo the *first* sound is from near surfaces, the *second*, or echo, from the distant surface.

9. SOUNDS IN LIQUIDS AND IN SOLIDS are more efficient

and more rapid than in air. Two stones rubbed together may be heard in water at half a mile. Cast iron conducts sound with ten and a half times the velocity of air; a string, or piece of deal, held to the ear, or between the teeth, having the vibrating body at the end, gives a vast increase of effect; and pipes convey sounds to immense distances. Solid bodies transmit sound to great lengths; the scratch of a pin at one end of a beam is heard at the other; and it is believed that a bar of iron ten miles long would transmit sounds almost instantaneously.

10. A beautiful experiment was lately instituted at Paris, to illustrate this fact, by Biot. At the extremity of a cylindrical tube, upwards of 3000 feet in length, a ring of metal was placed, of the same diameter as the aperture of the tube; and in the centre of this ring, in the mouth of the tube, was suspended a clock bell and hammer. The hammer was made to strike the ring and the bell at the same instant, so that the sound of the ring would be transmitted to the remote end of the tube through the conducting power of the matter of the tube itself; while the sound of the bell would be transmitted through the medium of the air included within the tube. The ear being then placed at the remote end of the tube, the sound of the ring, transmitted by the metal of the tube, was first distinctly heard; and after a short interval had elapsed, the sound of the bell, transmitted by the air in the tube, was heard. The result of several experiments was, that the metal of the tube conducted the sound with about ten and a half times the velocity with which it was conducted by the air, that is, at the rate of 11,865 feet per second.

11. THE STETHOSCOPE.—The fact of solids conveying sounds so much more perfectly than air, has been applied to the invention of an instrument called the stethoscope, or chest inspector. This is simply a wooden cylinder, one end of which is placed on the chest, and the ear applied to the other. This instrument becomes a means of ascertaining certain diseases in the chest almost as conveniently as if there were windows for visual inspection; for an ear familiar with the natural and healthy sound of the lungs instantly detects certain deviations from them.

12. In considering the means by which sound is communi-

cated, we may take for illustration a series of balls arranged in a line upon a table, or suspended together by threads. If at one end of the line we take a ball, and impel it with force against that which is next to it, the effect is observed at the opposite extremity of the line, and the ball there placed flies off from the rest, and leaves them almost stationary. Thus the intermediate balls serve merely to transmit the impulse from the one end to the other of the series. In the same manner it is that the agitation or impulse from which sound arises is transmitted to the air. This fluid, like every other body, consists of an infinite number of little particles; a single series of which may be represented to us by the balls in the above example. These particles are not even in contact with each other; they are separated by minute intervals, but are yet connected together by attractive and repulsive forces which tend to retain them perpetually in equilibrio. In every case, therefore, there is in reality a chain of such particles, reaching from the sounding body to the ear. The former, by its agitation, strikes that particle which is next to it, the intermediate ones seem to convey the impression, and the last one flying off strikes the sentient organ of hearing.

THE ANIMAL EAR.

13. The principles thus laid down respecting the origin and propagation of sound, and the causes of the different phenomena it exhibits, will be found to have parts applicable or corresponding to each in the structure of the human ear. Its mechanism appears to be exceedingly simple.

14. The human ear may be naturally considered as formed of three divisions; the *external ear, the tympanum*, and *the labyrinth*. The *external part of the ear* is called the *concha*, and is evidently adapted, like an ear-trumpet, to catch those pulses or vibrations of the air which we have already described as necessary to sound.

15. From the outer ear, or concha, we find a tube which leads into the head. The vibrations collected, and of course concentrated, at the bottom of this narrow tube fall upon a skin or membrane (11), called the *membranum tympani*, which is extended over the circular *opening* of the *bottom of the tube*. This is the *first* division of the ear.

16. The *second* division consists of that *cavity* which lies between the *membrane* (11), and a *circular opening* (6), called *the foramen ovale*. This *cavity* is called the *tympanum* on account of the likeness which, with its membrane spread over it, it bears to a *drum*. Behind the *membrane* we find a most curious chain of four little bones united by articulation and ligaments, and forming a *communication* between the membrane and the circular opening. These bones derive their names from the shape which they bear.

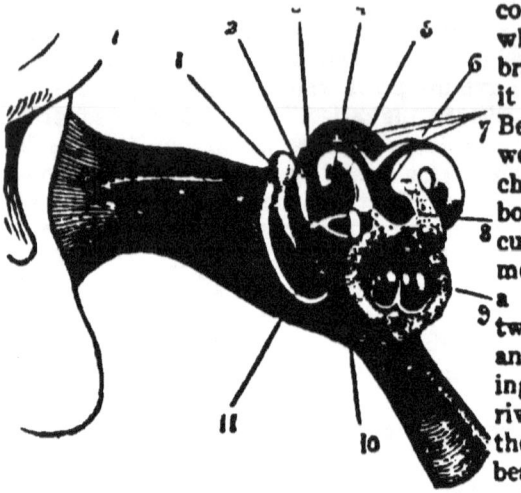

17. The *first* of them is called the *malleus* (1), in consequence of its resemblance to a hammer or mallet. The *round* part of it, which closely resembles the head of the thigh-bone, is called the *head of the malleus*. It has two legs or processes: *that leg which is attached to the centre of the membrane* is called the *handle* of the malleus. The *second* bone in the chain is named the *incus*(2), on account of its similarity to a blacksmith's anvil. It will be observed to have a depression, into which *the head of the malleus* is received. The next or third bone is called *os orbiculare* (3); it is no larger than a grain of sand: it is a sort of link which serves to connect the incus with the *stapes*(4), or last bone, which is so named on account of its similarity to a stirrup. This *concluding bone* in the chain plays upon a passage called the foramen ovale (6), which has been already mentioned. In addition to this series of bones, there is a communication between the drum or barrel of the ear and the *external air*, by means of a slender pipe called the Eustachian tube, which leads out of the tympanum into the back part of the mouth.

18. The third division of the ear is called the *labyrinth*. It

consists of the vestibule or middle cavity (8) of the semicircular canals (7), and of the *cochlea* (9), a portion so named from its similarity to a snail-shell. Over *this* complex *inner compartment*, which is filled *with water*, the nerve of hearing is spread as a lining. The action of the principal parts seems to be as follows:—The pulsations which we have spoken of are impressed upon the *membrane*, and cause it to *vibrate*. These *vibrations* are continued, through the agency of the *chain of bones*, to the *foramen ovale*, where the nerve is prepared to receive them, in agreement with the laws of hydraulic pressure, and are thus transmitted to the brain.

CHAPTER XXIV.
REFLECTION OF SOUND.

19. ANGLES OF INCIDENCE AND REFLECTION.—In sound or in light, the angle of incidence is equal to the angle of reflection (See Optics). When a wave of water strikes a wall it is thrown back with a degree of force proportionate to its mass. It is in the same manner that the pulses or waves of sound are reflected or thrown back from flat surfaces, thus producing what is termed an echo.

20. ECHOES are distinguished when the time between delivering a sound to its return is more than $\frac{1}{17}$ of a second, and as the sound goes and returns, to the speaker there can be no echo in less than $1\frac{14}{17}°$, or forty-seven feet, and syllables cannot be repeated in less than $\frac{1}{3}$ of a second more, or 161 feet for each syllable.

21. One of the most singular and distinctly marked illustrations of the reflection of sound forming a natural echo, occurs on the banks of the Rhine, near Lurley. Here a pistol being discharged, the projecting masses of rock on each side of the river multiplies it eight or nine times distinctly, and further through as many more reflecting points.

22. THE CONCENTRATION OF SOUND BY CONCAVE SURFACES produces many curious effects, both in nature and in art. A person uttering a whisper in one focus of an oval room, is very audible in the other. The whispering gallery of St. Paul's affords an example in illustration. The diagram represents the

circular dome; M shows the situation of the mouth of the speaker, and E that of the hearer. Now since sound radiates in all directions, a part of it will proceed from M to E, while

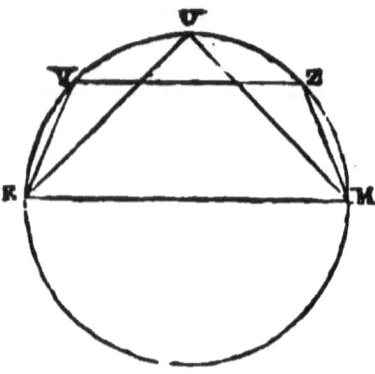

the other rays of it will proceed from M to U, and from M to Z, &c.; but the ray that impinges upon U will be reflected to E, while that which touches Z will be reflected to Y, and from thence to E; and so of all intermediate rays, which are omitted in the figure to avoid confusion. It is evident, therefore, that the sound at E will be much stronger than if it had proceeded immediately from M, without the assistance of the dome; for in that case the rays at Z and U would have proceeded in straight lines, and consequently could never have arrived at the point E. Hence the loudness of the whisper, and the thunder-like clap heard upon the shutting of the door, with which almost every one is familiar.

23. On the same principle is the reflection of sound from concave surfaces. Two inanimate busts may be made to appear as if holding a conversation, by placing them in the face of two large pasteboard mirrors arranged opposite to each other, when a whisper uttered from the one will seem to proceed from the other, by the reflection of sound.

24. Thus also a sea shell gives forth murmurs like the waves of the distinct sea, and a speaking trumpet is made to convey sound to a distance; for since sound radiates in all directions, it follows that if such radiation be prevented by confining it ir tubes, it may be carried to a great distance with very little diminution of effect, and hence the use and application of these trumpets, or tin speaking pipes, now commonly used for conveying intelligence from one part of a house to another. The trumpet used by persons partially deaf is in accordance with the same laws of sound.

25. Some amusing effects have been produced by operating

on sound with tubes and concave surfaces. What was termed the invisible girl, in which a voice was thought to proceed from a globe hung in the centre of a frame resembling a small bedstead, was a contrivance where the sound was conveyed through pipes ingeniously concealed, and opening opposite to the mouth of the trumpet, from which it seemed to proceed.

CHAPTER XXV.

SOUNDS DEPENDENT UPON THE NUMBER OF VIBRATIONS.—RELATIONS OF SOUND, MELODY, AND HARMONY.—TUNING FORK, MUSICAL GLASSES, KALEIDOPHONE.—MUSICAL RODS. —MUSICAL PLATES.

26. The difference of sounds, which depends on the different number of vibrations in the sounding body in a given time, divides them into those called *bass, low*, or *grave* notes, for comparatively few and slow vibrations; and those called *high, shrill*, or *sharp*, for vibrations more numerous and quick.

27. The frequency of vibrations on strings increases with their shortness, lightness, and tension; for if a string be long or heavy, there is a greater mass of matter to be moved, and hence a slower motion; and if a string be slack, the force of elasticity which pulls it from any deviation back to the straight line, is so much the less. It is found that a string taken of half the length, or of one-fourth the weight, or quadruple of the tension of another string, vibrates twice as fast on any one of these accounts.

28. When the number of impulses producing some continued sound has a simple relation, as of $\frac{1}{2}$, $\frac{1}{3}$, or $\frac{1}{4}$, &c., to the number producing some other sound which is heard, either simultaneously or a little before or after it, the ear is much and pleasingly affected, and the sounds are said to have a musical relation to each other, or to be *accordant*, while all others are termed *discordant*.

29. For instance, between two sounds, one of which beats twenty in a given time, and another ten, there will be a coinci-

dence at every *second* beat of the quicker; and between sounds whose beats are to each other as thirty is to twenty, there must be a coincidence in every third beat of the quicker, and so forth. Accordingly, all sounds which have such simple relations to each other are remarkably agreeable to the ear, while those whose coincident beats are further apart, are harsh and disagreeable. In order to comprehend the nature of *reciprocated vibration*, or *resonance*, let the reader keep in his remembrance the analogy between musical vibration and the oscillation of the pendulum, as explained at page 50. If he well understands the phenomena of the latter, he will readily comprehend those of the former. Galileo observed that a heavy pendulum might be put in motion by the least breath of the mouth, *provided the blasts were often repeated, and made to keep time exactly with the vibrations of the pendulum*—from the same sympathetic communication of vibrations will two pendulum clocks fixed to the same wall, or two watches lying upon the same table, take the same rate of going, though they would not agree with one another, if placed in separate apartments. Mr. Elliot indeed observed that the pendulum of one clock was even able to stop that of the other, and that the stopped pendulum, after a certain time, would resume its vibrations, and in its turn stop the vibrations of the other. We have here a correct explanation of the phenomena of *Resonance;* for the undulations excited by a vibratory body are themselves capable of putting in motion all bodies whose pulses are coincident with their own, and consequently with those of the primitive sounding body; hence the vibrations of a string when another tuned in unison with it is made to vibrate.

30. MELODY in music, is when notes having the same numerical relations above described, are played in *succession*. HARMONY is, when two or more such notes are sounded together.

31. MUSICAL NOTES, by whatever instruments produced, have to each other the same numerical relations in the beats or vibrations, which constitute them. The different qualities of tone, therefore, can only depend on the peculiarities of the single beats, as to whether they are sharp or soft, strong or weak.

32. That there might be correspondence in instruments

when played together, and a known pitch when played apart, it became necessary to fix on some tone or number of vibrations as a point of comparison. Hence *tuning-forks* have been made of steel, with length of prongs calculated to produce a certain note. This note is usually the fourth, A or *la*, from the base of the pianoforte, and vibrates about 430 times in a second; and when the note of the same name is heard in unison with this, the other notes can be easily adjusted according to the harmonic relations already explained.

33. If two vibrating tuning-forks, differing in pitch, be held over a closed tube, furnished with a movable piston, either sound may be made to predominate, by so altering the piston as to obtain the exact column of air which will reciprocate the required sound. The same result may be obtained by selecting two bottles (which may be tuned with water) each corresponding to the sound of a different tuning-fork; on bringing both tuning-forks to the mouth of each bottle, alternately, that sound only will be heard, in each case, which is reciprocated by the unisonant bottle; or, in other words, by that bottle which contains a column of air, susceptible of vibrating in unison with the fork.

34. MUSICAL GLASSES.—Glass vessels of different sizes being arranged as originally suggested by Dr. Arnott, are here shown. The small open circles represent the mouths of the glasses standing in a mahogany case, and the relation of the glasses to the written musical notes is shown by the common music lines and spaces which connect them. The learner discovers immediately that one row of the glasses produces the notes written
upon the lines, and the other row the notes between the lines, and he is mentally master of the instrument by simple inspection. This arrangement also renders the performance easy, for the notes most commonly sounded in succession are contiguous; and the relations of the notes forming a simple air are so obvious to the eye, that the theory of musical combination and accompaniment is learned at the same time. The

set of glasses here represented has two octaves, and the player stands at the side of the case, with the notes ascending towards the right hand, as in the piano-forte.

35. VIBRATION OF RODS.—If a rod be firmly fixed at one end, and allowed to vibrate freely through its whole length, tones of a very peculiar kind are found to result. Thus, a rod only two feet in length will give a tone as deep as that of the bell employed in the church of St. Paul's; and the Parisian clockmakers have availed themselves of this fact, in the construction of their ornamental chimney clocks, which by this means cost less, and strike without the sharp and dissonant tinkle common to light bells.

36. A very pretty instrument, called a "Kaleidophone," has 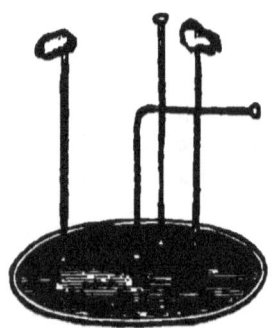 been contrived by Mr. Wheatsone, of which the accompanying cut is a representation. It consists of four vibrating rods, on which variously formed bodies are placed, and very beautiful and vivid figures are produced by merely drawing either of the rods out of the perpendicular, and then allowing them to vibrate freely.

37. VIBRATION OF PLATES. — The vibration of plates differs from those of rods in the same manner as the vibrations of membranes differ from those of cords, the vibrations of which cause the plate to bend in different directions, being combined with each other, and sometimes occasioning singular modifications. These vibrations may be traced through wonderful varieties by Professor Chaldni's method of strewing dry sand on the plates, which, when they are caused to vibrate by the operation of a bow, is collected into such lines as indicate those parts which remain either perfectly or very nearly at rest during the vibrations. Dr. Hooke had employed a similar method for showing the nature of the vibrations of a bell, and it has sometimes been usual in military mining to strew sand on a drum, and to judge, by the form in which it arranges itself, of the quarter from which the tremors produced by countermining proceed.

MUSICAL INSTRUMENTS—THE ORGAN.

38. It usually happens that the vibration of a cord deviates from the plane of its first direction, and becomes a rotation or revolution which may be considered as composed of various vibrations in different planes, and which is often exceedingly complicated. We may observe this by a microscopic inspection of any luminous point on the surface of the cord; for instance, the reflection of a candle in the coil of a fine wire wound round it. The velocity of the motion is such, that the path of the luminous point is marked by a line of light, in the same manner as when a burning coal is whirled round; and the figures thus described are not only different at different parts of the same cord, but they often pass through an amusing variety of forms during the progress of the vibration; they also vary considerably according to the mode in which that vibration is excited.

OF SOUNDS PRODUCED BY MUSICAL INSTRUMENTS.

Organ.

39. Musical instruments may be divided into three distinct classes, viz. :—

1st. Stringed or membranous instruments, as the harp, drum, &c.

2nd. Wind instruments, as the flute, French horn, basoon, organ, &c.

3rd. Metallic instruments, as cymbals, bells, musical boxes, Jews' harps, &c.

40. (1.) The action of the hand upon the harp, the bow upon the violin, the fingers upon the tambourine, or the sticks upon

Drum.

Harp.

the drum, all cause vibration. The sound produced does not, in the first instance, arise from these vibrations acting upon the external air, but in the constriction of the atmospheric particles interspersed within the elastic animal fibre, and these particles thence expanding into the surrounding air originate the production and diffusion of sound. Where the elasticity is destroyed by damp, motion is produced without occasioning sound, as in a wet drum head.

Violin and Bow.

Basoon.

41. (2.) Wind instruments may be said to consist of three kinds :—

1st. The clarionet, flute, &c., into which the air is propelled in nearly an equable manner, and then harmony is occasioned

FLUTE—HORN—PIANOFORTE.

by opening and closing certain apertures in their sides, which may be considered as alterations in their structure.

Clarionet and Flute.

Serpent.

2nd. In iustruments admitting of no alteration of structure, as the trumpet, French horn, &c., the harmony is propelled by the mouth and lips, the metallic surfaces merely deepening its intonation; in organs, each tube is constructed to utter a separate note, and the harmony is produced by opening or shutting the tubes at pleasure.

Horn.

Guitar.

Piano.

3rd. Metallic instruments, as bells, are dependent upon their composition and relative size for their diversity of tone, and the notes of musical boxes are regulated by the length and thickness of their springs, which being elevated by points fixed in a moveable barrel, produce in their rebound the degree of harmony required.

Cymbals and Triangle.

Lyre.

42. In all these instances it will be observed the same principle prevails of a constrictile power being first produced, which afterwards leads to that modified expansion which creates *sound, tone, and harmony.*

43. The pitch or tone of a tubular wind instrument, just as a musical string, has relation to its length and the vibrations causing the sound. The waves or condensation of air passing from the mouth to the extremity of the tube, being more frequent, therefore, as the tube is shorter. When the bottom of the tube is closed the wind has to come back again, and thus renders the note twice as grave.

44. THE COMPASS OF VARIOUS INSTRUMENTS is as follow:—The *harp* through six octaves, from five lines below the bass clef to six lines above the treble; the *piano* through *five* octaves, and with extra keys to *six*; the guitar through two and a quarter; the *clarionet* three and a half; the *horn* three; the *basoon* three; the *flute* three; the *violin* two and a half; the *violincello* two and a quarter; *human voices* two, the *soprano* two notes below the treble to three above, the *tenor, middle tone* of bass to middle of treble, the *bass,* note below the line of bass clef to the second line of treble.

OPTICS.

CHAPTER XXVI.

THEORIES AND PHENOMENA OF LIGHT.—LIGHT AS AN EFFECT.—VELOCITY OF LIGHT.

1. OPTICS is the SCIENCE OF LIGHT AND VISION, and is divided into three parts:—1, DIOPTRICS, which explains the properties of *refracted* light; 2, CATOPTRICS, which investigates the properties of *reflected* light; 3, PERSPECTIVE, which relates to drawing and painting, or the representation of objects on a plane surface.

2. THEORIES OF LIGHT.—Concerning the nature of LIGHT,

two theories are at present very ably maintained by their respective advocates: one is termed the NEWTONIAN or *molecular*, the other the HUYGENIAN or *undulatory* theory; each of which is signally ingenious and beautiful in detail, and alike supported by men of the profoundest mathematical skill.

3. PHENOMENA OF LIGHT.—The *Newtonian* or molecular theory of light, considers it to consist of unconceivably small particles emanating from the sun or any other luminous body. According to *Huygens*, it consists in the undulations of a highly elastic and subtle fluid pervading all bodies, and propagated round the luminous centre in *spherical waves*, like those arising in a placid lake when a stone is dropped into the water. This theory is now adopted by several distinguished opticians.

4. But whichever of these theories be adopted, LIGHT is an emanation from the sun and other luminous bodies, and, being projected from them to the eye, renders them *visible*. Its absence is called *darkness*. It moves with great *velocity*, and in *straight lines* where there is no obstacle, leaving *shadows* where it cannot fall. It passes readily through some bodies, which are therefore called *transparent*; but when it enters or leaves their surfaces obliquely, it suffers at them a degree of *bending* or *refraction* proportioned to the obliquity; and a beam of white light thus refracted or bent, does not all bend equally, but is divided or resolved into beams of what are called the *elementary colours*, which colours, on being again blended, become the white light as before.

5. LIGHT AS AN EFFECT.—Thus we know little of light, except by the effect which it produces. Light follows, indeed, the law of all forces or influences emanating from a centre, for the intensity or degree of light *decreases* as the *square of the distance* from the luminous body *increases*.

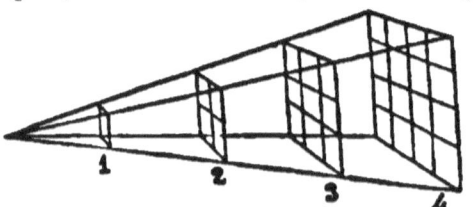

Thus at the distance of two yards from a candle, we shall have four times less light than we should have if we were only one yard from it, and so on in the same proportion.

6. RAYS OF LIGHT.—Light consists of separate parts or atoms called *rays*, which are independent of each other. These are projected from the luminous body in straight lines, which is seen when the sun darts his beams through a cloud of smoke or dust. It is also proved by the fact, that we cannot perceive objects through a bent tube. If light be admitted into a dark room by a small hole in the shutter it illuminates a spot in the room exactly opposite the shutter. The smallest portion of light so admitted to a room is called a *ray* of light, and all quantities of light are bundles of these rays. Light is called *homogeneous* when resolved into its seven primitives, and *hetereogeneous* when the seven being united it appears white. A BEAM of light is a body of parallel rays, and a *pencil* of rays is a body of *diverging* or *converging* rays. The point at which converging rays meet is called the Focus.

7. VELOCITY OF LIGHT.—The particles or vibrations of light travel twelve millions of miles in a minute. They therefore take eight minutes in coming from the sun to us. The rapidity of the motion of light was discovered by M. Roemer, who observed that the eclipses of Jupiter's satellites took place about sixteen minutes later at that part of the earth's orbit furthest from the planet; as the earth's orbit is 190 millions of miles in diameter, it gave the rapidity with which light travelled very accurately.

8. LIGHT THE EFFECT OF VARIOUS EXCITEMENTS.—Light, independent of its transmission from the sun and other luminous bodies, is also the effect of various excitements. Friction produces it, and phosphorus produces it. Snow, the diamond, and the Bologna stone, appear to absorb and radiate it. Some combinations evolve it, and some plants give flashes. Rubbing the eyes in the dark, and also their inflammation, produces flashes of light as well as heat, and crystallization is accompanied by flashes of light.

9. The bodies which emit rays of light, as the sun and fixed stars, are called *luminous*. Those that transmit the rays of light, such as air, water, glass, crystal, &c., are called *pellucid* or *transparent*. Those that reflect a part, and suffer a part only to pass through them, are called *semi-transparent;* and substances which do not suffer the rays of light to pass through them, are called *opaque*.

DIOPTRICS.

CHAPTER XXVII.

REFRACTION.—OF THE PRISM.—REFRACTIVE POWERS OF MEDIA, MULTIPLYING GLASS, &c.

10. The term Dioptrics signifies to *see through*, and it accordingly denotes that branch of optics which treats of the transmission of the rays of light through transparent bodies.

REFRACTION.—Things which suffer the rays of light to pass

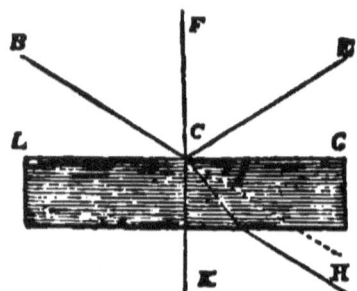

through them, such as water or glass, are called *media*. When rays of light enter these, they do not proceed in a direct line, but are said to be *refracted*, that is, bent out of their course, as seen in the drawing. The ray of light proceeding from B through the glass L G, is bent from the point C, instead of passing in the direction of the dotted line. But if the ray fall perpendicularly on the glass at F C, there is no refraction, but the ray proceeds in its passage through the glass in a direct line as F K ; hence refraction only takes place when the rays fall obliquely or aslant on the media.

11. When a ray of light leaves the medium in which it has been refracted, and enters that through which it was originally passing, it is refracted as much in a contrary direction, so that it moves *exactly parallel to what it did before entering the glass*, as shown in the line B C H.

12. The refraction or bending of light is easily understood by means of a prism or triangular piece of glass, as *b c*. A ray from *a* falling on the surface at *b*, is bent towards the internal perpendicular, and, therefore, reaches *c* ; but on escaping

REFRACTION OF LIGHT. 129

again at *c*, it is bent away from the external perpendicular, and thus, with its original deviation doubled, goes on to *d*.

13. If a coin be placed in a basin, so that on standing at a certain distance, it be just hid from the eye of an observer by the rim or edge of the basin, and then *water* be poured in by a second person, the first keeping his position ; as the water rises, the coin will become visible, and will appear to have moved from the side to the middle of the vessel.

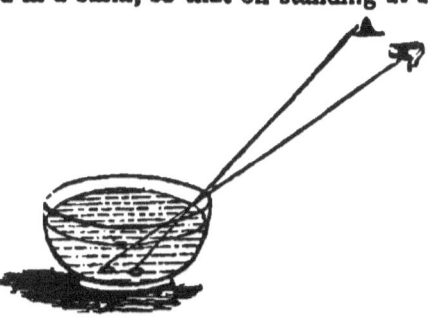

14. The above facts lead to an important axiom in optics, namely, *That we see every thing in the direction of that line in which the rays approach us last.* Hence the sun is seen a lit-

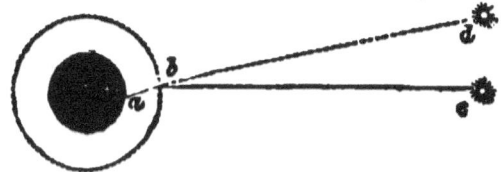

tle time before he comes to the horizon, and as long after he sinks beneath it in the evening, because the rays of the sun strike upon the atmosphere at a height of forty-five miles before the sun himself is above the horizon, and are by refraction bent towards the earth, so as to a spectator at *a* supposed on the surface of the earth, the sun really at *c* appears to be higher, because its ray on reaching the atmosphere at *b* is bent downwards, and the ray coming horizontally from *b* to *a*, appears to come from c. This applies only to oblique rays: when the sun is in the zenith there is no refraction.

15. REFRACTIVE POWERS OF MEDIA.—As regards the refractive powers of transparent substances, the general rule (with some limitations) is, that it is proportionate to their different densities; that is, the denser medium has the greatest refracting power. When a ray of light passes from air into water, the refraction is as 4 is to 3; but when it passes from air into glass, it is as 3 is to 2, that is, the measures of the ratios are $\frac{4}{3}$ and $\frac{3}{2}$, multiply both fractions for any number, as 12, and the latter will be seen to be the larger.

16. Hence refraction increases from the most perfect vacuum that can be formed, through *air, fresh water, salt water, glass*, and so on. It was from the great refractive powers of the diamond and of water that Newton inferred that they contained inflammable substances, which chemistry has since proved to be the case, from the discovery of hydrogen as the principle constituent of water, and of the diamond to be pure carbon. (See Chemistry.)

17. In looking through what is called a multiplying glass, we have also an illustration of the effects of refraction: for if a small object, such as a fly, be placed at d, an eye at e will see

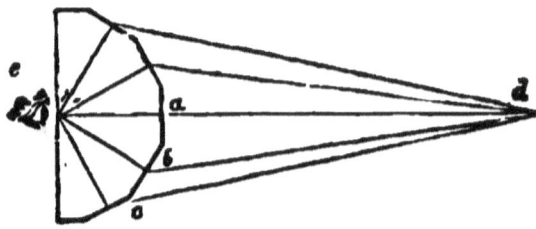

as many flies as there are distinct surfaces or fascets on the glass; for first the ray $d\,a$ passing perpendicularly, and, therefore, straight through, will form an image as if no glass intervened; then the rays from d to the surface b will be bent by the oblique surface, and will show the object as if it were in the direction $c\,b$, and the light still falling on the more oblique surface c, will be still more bent, and will reach the eye in the direction of the bent lines, exhibiting a similar object also in that direction, and so of all the other surfaces.

18. Owing to the bending of light in passing through media of different densities, a beautiful phenomenon is often observable in a day of warm sunshine. Black or dark coloured substances, by absorbing much heat and light from the sun's rays, and warming the air in contact with them, until it dilates and rises in the surrounding air as oil rises in water, cause the light from more distant objects, reaching the eye through the more rarified medium, to be bent a little, and owing to their rising irregularly from the influence of the wind and other causes, these objects acquire the appearance of having a tremulous motion. Thus on a warm clear day the whole landscape at times appears to be dancing. The same phenomenon is often observed on board a steam-boat, by looking through the heated air near the chimney.

19. MIRAGE, &c.—The elevation of coasts, ships, and mountains, above their usual level, when seen in the distant horizon, has been long known, and are to be ascribed to *refraction*. The name *mirage* has been applied by the French to the same class of phenomena; and the appellation *fata morgana* has been given by the Italians to similar appearances, which have been repeatedly seen in the Straits of Messina. When the rising sun throws his rays at an angle of 45° on the sea of Reggio, and neither wind nor rain ruffle the smooth surface of the water in the bay, the spectator on an eminence in the city, who places his back to the sun and his face to the sea, observes, as it were upon its surface, a numberless series of pilasters, arches, and castles, distinctly delineated; regular columns, lofty towers, superb palaces, with balconies and windows; extended vallies of trees, delightful plains with herds and flocks; armies of men on foot and horseback, and many other strange figures, in their natural colours and proper actions, passing one another in rapid succession.

CATOPTRICS.

Argand Lamp.

CHAPTER XXVIII.

LAWS OF REFLECTION.—ANGLE OF INCIDENCE AND REFLECTION.—REFLECTION FROM CONCAVE AND CONVEX SURFACES.—MIRRORS.—FOCI. —CARRIAGE LAMPS.

20. The word CATOPTRICS is composed of two Greek words, one of which signifies *from* or *against*, and the other to *see*, hence it is used to designate that branch of the science of optics which treats of *light sent back or reflected* from plane or spherical surfaces, and includes the principles upon which mirrors or specula are constructed.

21. When a marble is shot against a wall as from a tube, it will be returned in the same line, and is said to *rebound*. When a ray of light strikes an opaque body at a right angle it flies back in a similar manner, and is, in optical language, said to be *reflected*.

22. If a marble be shot sideways at a hard body, as from A to C, instead of its coming back in the same line, it will, after striking the plane surface, go off to the opposite corner to B. Light obeys the same law, and if it fall against a mirror, or any other reflecting body, it is then said to be *reflected* at a certain *angle*.

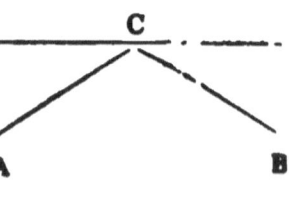

LAWS OF REFLECTION.

23. From this is deduced one of the principal laws in optics, viz., That the angle of *reflection* is always equal to the angle of *incidence*. The line which a ray of light describes in going to an object is called the line of *incidence*, and the line in which it rebounds or comes back, is called the line of *direction*. The surface from which a ray of light is sent back is called the *reflecting surface*.

24. This rule we will first apply to plane surfaces. Let A B be the looking-glass, and D the spectator opposite to it. The ray from his eye will be reflected in the same line A a, but the ray D B 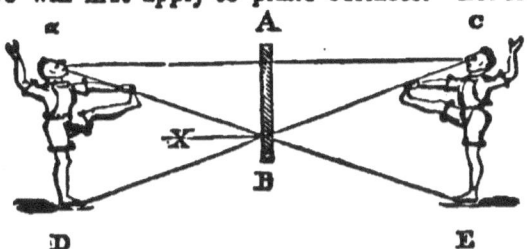 coming from his foot in order to be seen at the eye, must be reflected by the line B a ; for if X B be a line perpendicular to the glass, the incident angle will be D B X, equal to the reflected angle a B X; and, therefore, the foot will appear behind the glass at E, along the line a B E, because that is the line in which the ray last approached the eye. That part of the glass A B, intercepted by the lines A C and B E, is equal exactly to half the length C E or a D. a A B and A B C may be supposed to form two triangles, the sides of which always bear a fixed proportion to one another, and if a C is double of a A, as in this case it is, C E will be double of A B, or at least of that part of the glass intercepted by a A C and D B E. Hence if a person wishes to see his complete image in the looking-glass, it is not necessary that the glass should be as high as he, but half that height will suffice.

25. If a person looks at the reflection of a candle in a looking-glass he will see two images, one much fainter than the other. The same may be observed of any object that is strongly illuminated, and the reason of the double image is, that a part of the rays are immediately reflected from the upper surface of the glass, which forms the faint image, while the greater part of these are reflected from the further surface or silvered part, and forms the vivid image. To see these two

images the man must stand a little sideways, and not directly before the glass.

26. REFLECTION FROM CONCAVE AND CONVEX SURFACES

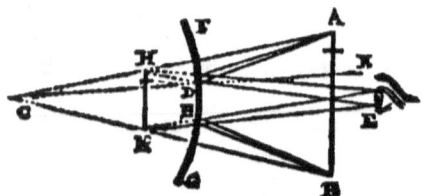
Fig. 1.

—The same law of reflection applies to concave and convex surfaces. Thus in the convex mirror F G, the rays which fall from the top of the cross at A to the mirror at D, fig. 1, are reflected to the eye at E at exactly the same angle, which thus receives them as if they came from a real object at H. The rays which fall from B are reflected in the same way, and enter the eye as if they proceeded from K. Precisely similar the rays which fall from M and N upon the concave mirror F G at D, fig. 2, are reflected to the eye at E, and enter it as if they had come from a real object at O P; the angle C D E being equal as before to the angle C H N.

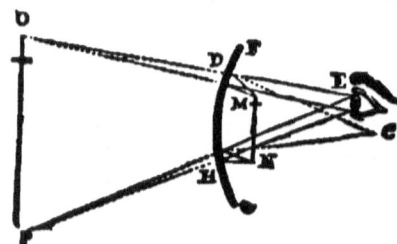
Fig. 2.

27. When parallel rays $g\,c\,l$ fall upon a concave mirror A B, they will be made to converge, or meet at a certain point, at m, at half the distance of the surface of the mirror from c, the centre of its concavity. Thus let e be the centre of the mirror A B, and let the parallel rays reflected back from that mirror, fall upon it at the points $d\,e\,f$. Draw the lines $c\,i\,d$, $c\,m\,e$, and $c\,h\,f$ from the centre c to these points, and all these lines will be perpendicular to the surface of the mirror because they proceed to it like so many rays from a centre.

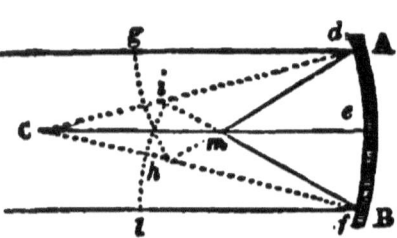

Make the angle cdh equal to the angle $\bf{z}\,dg$, and draw the line dmh, which will be the direction of the ray g, after it is reflected from the point of the mirror, so that the angle of incidence, gdi is equal to the angle of reflection hdi, the rays making equal angles with the perpendicular on its opposite sides.

28. Draw also the perpendicular ckf to the point $,f$, where the ray lf touches the mirror f. Having made the angle cfi equal to the angle cfl, draw the line fmi, which will be the course of the ray elc after it is reflected from the mirror.

The ray cme which ought to pass through the centre of the mirror and fall upon it at e, perpendicular to the surface, is reflected back from it in the same line emc, because the mirror is not transparent.

29. All these reflected rays meet at the point m, and in that point the image of the body, which emits the parallel rays gcl will be formed, which point is distant from the mirror equal to half the radius emc of its convexity.

30. Thus in all concave mirrors, of whatever substance they may be formed, the focal distance is exactly equal to one-half of the radius of the mirror's concavity. The *focus* or *fireplace*, where the rays meet at a point, is so called on account of these collected rays possessing the power of burning any combustible body placed there. Thus if two large mirrors N and B be

placed opposite to each other at the distance of several feet, and red hot charcoal be put in the focus D, and some gunpowder in the other focus, it will presently take fire. This experiment may be varied by placing a thermometer in one focus, and lighted charcoal in the other, when it will be seen that the quicksilver will rise as the fire increases, though another thermometer, at the same distance from the fire, but not in the focus of the glass, will fail to be affected by it.

31. If an object A B C, a bottle half filled with water, be held before a concave mirror, the image of the bottle will be seen inverted, but the mind will conceive the water to be on that part of the bottle which appears to be lower. If the cork be now taken out, and the mouth of the bottle be held downwards, as the water runs out the illusion of the mind will be such that the bottle will seem to *fill*, till all the water be gone.

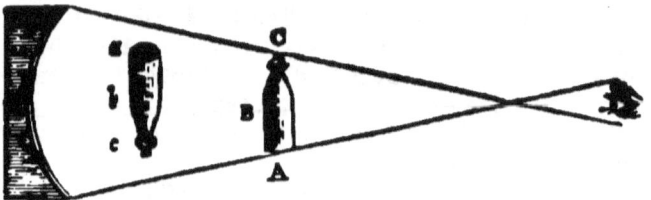

32. The appearance of the image in the air, between the mirror and the object, has been productive of many agreeable deceptions, which, when exhibited with art and an air of mystery, have been considered as something almost miraculous. If a partition be made between two rooms as A, having a small square hole in the centre B, and in the back room a concave mirror E F be placed, with an inverted figure C. This figure will be seen by the eye at G, as standing immediately before

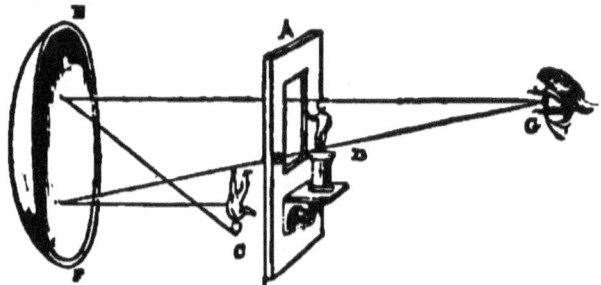

the square. The course of the line D shows the direction of the rays. Sometimes a spectral appearance has been exhibited by the same means, which vanishes upon the approach of the spectator.

33. If a burning candle or lamp be placed nearer to a concave mirror than its focus, its rays will enlighten a larger space. Hence may be understood the construction of carriage and other lamps. Concave mirrors are chiefly used in reflecting telescopes, and the properties of these mirrors are very important. Images reflected from them appear larger than the real objects, provided these objects are within the focus, and the image of an object receding continually from a concave mirror, becomes continually greater, provided it does not recede beyond the focus, in which case it will appear less and inverted, as already explained.

CHAPTER XXIX.

DEFINITION OF LENSES.

34. A LENS is a piece of glass ground into such a form as to collect or disperse the rays of light which fall through it.

35. Lenses are of various kinds, such as plano-convex, plano-concave, double convex, double concave, and, if concave on one side and convex on the other, it is called a meniscus.

A plano-convex lens, shown at A, is a lens having one of its surfaces convex and the other plane.

A plano-concave lens, represented at B, is a lens, one of whose surfaces is concave and the other plane.

A double convex lens, shown at C, is a solid formed by two convex spherical surfaces, having their centres on opposite sides of the lens. When the radii of its two surfaces are equal, it is said to be equally convex; and when the radii are unequal, it is said to be an unequally convex lens.

A double concave lens, as shown at D, is a solid bounded by two concave spherical surfaces, and may be equally or unequally concave.

A meniscus, shown at E, is a lens, one of whose surfaces is convex and the other concave, and in which the surfaces meet if continued. When the convexity exceeds the concavity, it may be regarded as a convex lens. (It is called a meniscus, because it resembles the crescent moon.)

A line passing through the centre of a lens, as at F, is called its axis.

A spherical lens is a sphere, all the points in its surface being equally distant from the centre.

CHAPTER XXX.

REFRACTION OF LIGHT BY PRISMS AND LENSES.

36. When the rays of light proceed towards each other, as in

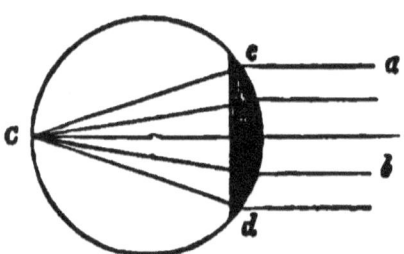

the subjoined figure from $c\,d$ to C, they are said to *converge;* when, as from the point C, they spread to c and d, they are said to *diverge.*

37. Parallel rays falling upon a plano-convex lens (as in the figure, where the lens is represented by the shaded part) meet at a point behind it, the distance of which is *exactly equal to the diameter of the sphere,* of which the lens is a portion.

38. Rays of light passing through a double convex lens, A B D E, will have the distance of the focus of parallel rays,

only equal to the *radius of the sphere*, that is, of half its diameter at *f*.

39. Convex lenses render the rays passing through them *converg-ent*. Concave lenses render them *divergent*. The power of the heat collected in this focus is in proportion to the common heat of the sun, as the area of the glass is to the area of the focus, and may be 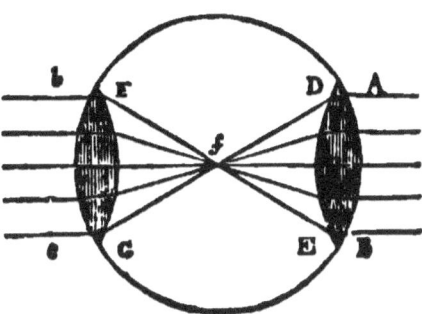 a hundred or even a thousand times greater in the one case than in the other.

40. When parallel rays, *a b c d e*, fall upon a double *concave* lens, A B, and pass through it, they will diverge, and the degree of diverging will be precisely so much as if they had come from the point, C, which is the centre of the concavity 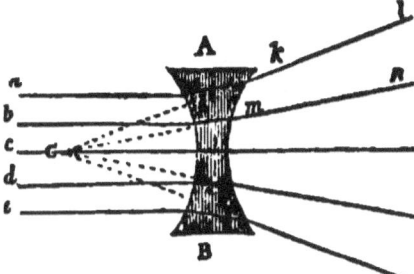 of the glass. This point, C, is called the virtual or imaginary focus.

41. If the lens had been a plano-concave, as the lens B, the rays would have diverged after passing through it, as if they had come from a radiant point at the *distance* of a *whole diameter* of the convexity of the lens.

42. Thus, the ray *a*, after passing through the glass A B, will go on in the direction *k l*, as if it had proceeded from the point C, and no glass had been in the way. The ray *b* will go on in the direction *m n*, &c. The ray C, falling directly upon the middle of the glass, suffers no refraction in passing through it, but goes on in the same rectilinear direction, as if no glass had been in the way.

43. If rays come more convergent to such a glass, they

will continue to converge after passing through it; but they will not meet so soon as if no glass had been in the way, and they will incline towards the same side to which they would have diverged, if they had come parallel to the glass.

44. Images formed by a *concave* lens, or those formed by a *convex* lens, where the object is *within* its principal focus, are in the same position with the objects they represent; they are also *imaginary*, for the refracted rays never meet at the foci whence they seem to diverge. The images of objects, however, placed beyond the focus of a convex lens, are inverted and *real*; for the refracted rays meet at their proper foci.

CHAPTER XXXI.

OF COLOURS.—DECOMPOSITION OF LIGHT BY THE PRISM.—BY ABSORPTION.—HEATING POWER OF THE SPECTRUM.—ILLUMINATING POWER.—CHEMICAL EFFECT.

45. Light is not a simple but a compound substance, and is susceptible of decomposition into its elementary parts. They

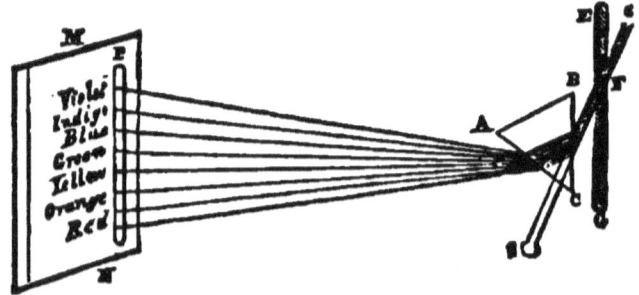

are the primitive colours, *red, blue,* and *yellow,* forming in their combination, when thrown on a *spectrum,* four other colours, *orange, green, indigo,* and *violet.*

46. To decompose a ray of light, it is only necessary to place

a prism in the situation as shown in the diagram A B C. The shaded perpendicular E G represents the section of a shutter, in which a hole is cut at F to let in a beam of light S S, which, falling on the prism, suffers different degrees of refraction, and is separated into the seven prismatic colours, as shown in the opposite wall at M.

47. Here we may observe, that the red ray being the least refrangible is least turned from its course, and accordingly occupies the lowest position on the spectrum, the coloured surface being so called at P N. The other colours are marked in the order of their refrangibility. If this spectrum be divided into 360 parts, the red will occupy 45 of them, the orange 27, the yellow 48, the green and the blue 60 each, the indigo 40, and the violet 80.

48. If these colours, in the same proportion, be painted on the side of a circular piece of board, dark in the centre and lighter at their edges, till they blend with those adjoining, and the board be made to revolve on a pivot, the painted surfaces will appear white.

49. The decomposition of light by absorption has been investigated by Sir D. Brewster. If a beam of white light be transmitted through a coloured prism, the latter absorbs a portion of the former. By looking at the spectrum P through the prism A B C, a prism of blue glass being interposed between the eye and the spectrum, Sir David found that the blue glass had absorbed the *red* light, which, when mixed with the *yellow*, constituted the parts of the green space next to the yellow. Thus, by absorption, *green* light was decomposed into *yellow* and *blue*, and orange light into yellow and red.

50. The conclusions drawn from these phenomena were, that the solar spectrum consists of three spectrums of equal lengths, viz., a *red* spectrum, a *yellow* spectrum, and a *blue* spectrum. The primary red spectrum has its greatest intensity in the middle of the *red* space in the solar spectrum. The primary *yellow* spectrum has its maximum in the middle of the *yellow* space, and the primary *blue* spectrum has its maximum between the blue and the indigo space. The two minima of each of the three primary spectra coincide at the two extremities of the solar spectrum. Thus then it is concluded, that certain proportions of *red*, *yellow*, and *blue*, form WHITE, and that

these exist at every point of the solar spectrum constituting its three PRIMARY COLOURS.

51. PHYSICAL PROPERTIES OF THE SPECTRUM.—HEATING POWER.—It was discovered by Sir William Herschel that the heating power of the spectrum gradually increased from the violet to the red extremity, and even beyond it. Hence he concluded that there are invisible rays in the light of the sun which had the power of producing heat, and which had a less degree of refrangibility than red light. Sir Henry Englefield confirmed his results, and obtained the following measures:—

	Temperature.		Temperature.
Blue	56°	Red	72°
Green	58	Beyond red	79
Yellow	62		

The place of maximum heat has recently been found to vary with the substance of which the prism is formed. Thus, in water, alcohol, and oil of turpentine, it is in the yellow; in crown glass, in the middle of the red; and in flint glass, beyond the red. In other substances it is intermediate between these two points.

52. ILLUMINATING POWER.—M. Fraunhofer, a celebrated philosopher of Munich, discovered, by means of a photometer, or measurer of the intensity of light, that the most luminous rays of the spectrum are not situated in the middle, but nearer the red than the violet end, in the proportion of 1 to 4 nearly, and that the mean ray is almost in the middle of the blue space. The same philosopher also discovered that the spectrum is covered with dark and coloured lines parallel to one another, and perpendicular to the length of the spectrum. They have always the same position in the coloured spaces in which they are found, their proportional distances varying with the nature of the prism by which they are produced. Their number, however, their order, and their intensity, remain invariable, provided light of the sun or moon be employed. One of the most important practical results of the discovery of these fixed lines, is, that they enable philosophers to take the most accurate measures of the refractive and dispersive powers of bodies and by measuring the distances of the lines, their discoverer

computed a table of the indices of refraction of different substances.

53. CHEMICAL EFFECT OF THE SPECTRUM.—The spectrum exercises a chemical influence on certain bodies. The effect, for instance, produced on nitrate of silver, varies with the nature of the coloured space where it is placed, and other substances are similarly affected. The solar rays possess also a magnetising power. If the violet rays be collected in the focus of a convex lens, and this focus carried from the middle of one-half of a small needle to the extremities of that half without touching the other, it will acquire perfect polarity. The indigo, blue, and green rays produce this effect, but the others do not. Exposure to the sun's rays, under peculiar circumstances, can be also made to produce similar results on certain bodies.

54. The most marked chemical effect of solar light, is its power of darkening the white nitrate of silver; and M. Daguerre has, by a long course of experiments, reduced to practice the means of *fixing* the images of objects produced by the camera obscura or solar microscope.* The means are a skilful modification of nitrate of silver to the surface of a metallic plate.

55. THE COLOUR OF BODIES arises from the absorption of some rays of the spectrum, and the reflection of the others. Transformations of visual colours are as follow :—A square of red long viewed produces a light green border, and afterwards a square of light green; white produces black, and black white; red *blue*; purple *green*; blue *yellow*; green *red*. Kirk has found that colours possess different powers of imbibing odours. Thus black absorbs more than white, and other colours various degrees.

56. In the prismatic spectrum, violet rays indicate heat as 1, green as 4, yellow as 8, and red as 16. Some philosophers ascribe colours simply to various degrees of intensity. The different colours of very remote stars prove that the causes of colour are universal. Late experiments prove that the green colour of plants arises from nitrogen, red from oxygen, and indigo and violet from hydrogen. The intensity of light is in red 2, in orange 30, in yellow and green 100, in the blue 32, in the indigo 18, and violet 3. The heating powers of the red rays in the spectrum to green is 55 to 26, and to voilet 55 to 16.

* See ILLUSTRATION Daguerreotype.

CHAPTER XXXII.

THE RAINBOW.

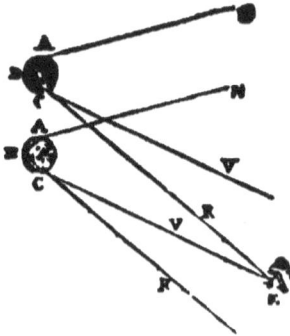

57. PRIMARY BOW.—This meteor in the heavens depends upon the refractive power of the drops of rain, and is a natural instance of the decomposition of the sun's light. If A B C be a drop of rain, and S A a ray of the sun falling upon the upper side of it, that ray will be refracted to B, reflected to C, and in going out of the drop will be separated into the primary colours.

58. The red ray being the least refrangible will take the lowest course, as C R, while the violet being most refrangible will take the highest course, as C V.

The intermediate colours between these according to their different degrees of refrangibility.

59. The sun's light being refracted and reflected in the same manner by the drop (7) the red light from this drop will take the course C R, parallel to C R from the drop (1), while the violet light will take the course C V, parallel also to C V in the drop (1). It is evident then that the eye at E will receive the violet ray *only* from the drop (7), all the other colours falling below it.

60. If then the eye receives the red light from the highest drop in the direction C R E, and the violet light from the lowest drop in the direction C V E, it is easy to conceive that it will receive the intermediate colours of *orange, yellow, green, olive* and *indigo*, from the intermediate drops in the order, *red, orange, yellow, green, blue, indigo*.

61. It may be asked why, if the rainbow depends on drops of falling rain, does it not immediately disappear as the rain reaches the ground? This is really the case. We do not see the same rainbow for a second; for immediately the drop that reflects the light, passes the spot on which the eye is fixed, we no longer perceive the rays coming from it, but we have a continued impression of the rainbow, because as fast as one drop passes away another supplies its place, and we do not see any intermission in the colours reaching the eye; as an impression remains on the retina for the sixth part of a second only, and the reflection from each drop comes more quickly than this.

62. SECONDARY BOW.—Another prismatic bow is generally seen above the rainbow, but of fainter colours. It also differs from the primary by its colours being in a reversed order. In the secondary bow the ray S A (1) will be refracted to B, reflected to C, thence to D, and upon its emergence at D, each colour will take its course correspondent to its degree of refrangibility, the red in the direction D R, and the violet being most refrangible, in the direction D V E; an eye at E, there.

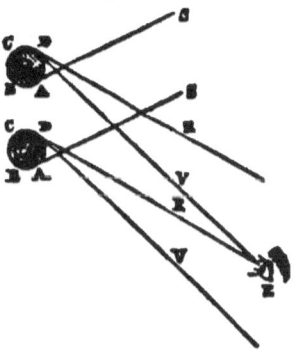

fore, will receive the red light only from the drop (1), all the other colours falling above it.

63. In the formation of the *primary* bow there are *two refractions* and *one reflection*. In the secondary there are two refractions and two reflections.

64. The angle made by the violet rays with the incident ones is 54° 0, and that of the red rays 50° 58'. At every reflection many rays pass out of the drop without being reflected, thus the secondary bow, which is produced after two reflections, is formed by fewer rays than the first, which is produced by one reflection: hence its faint appearance.

THE EYE.

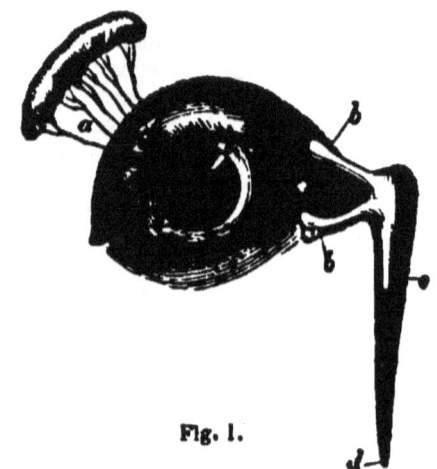

Fig. 1.

CHAPTER XXXIII.

65. THE HUMAN EYE is an optical instrument of the most perfect construction. It is a kind of telescope, or rather a camera obscura. It is of a globular shape, rather more prominent before than behind, and is composed of three coats or

skins, one covering the other like the coats of an onion. These inclose *three humours* hereafter to be explained. It is placed in a bony cavity, formed of seven bones, called the orbit; but, as bone is a hard substance, and would be liable to hurt so tender an organ in its motions, a soft cushion has been provided for it to repose on: the orbit is lined with fat, which not only forms a soft bed for the eye to rest upon, but facilitates its various motions. The eye is placed in the highest part of our bodies, and thus the better disposed for looking around, just as a spectator would endeavour to obtain the highest place, to view any passing spectacle.

66. To defend the eyes from too much light, and to prevent their being incommoded by any substances that might slide down upon the forehead, and thence fall into them, the *eyebrows* are placed above, which also add greatly to the beauty of the eyes.

67. The eyelids also, like two substantial veils, protect and cover the eyes when we are asleep. When we are awake, they diffuse by their motion, and by a peculiar secreting organ, a fluid over the eye, which cleans and polishes it, and thus renders it fitter for transmitting the rays of light.

68. The apparatus for this purpose may be seen in the cut. At fig. 1, *a*, is what anatomists call the lachrymal gland, which secretes (makes) the tears, or moisture for washing the eye, and keeping it moist. The little hollow pipes, or tubes, running from it towards the eye, are for the purpose of conveying to it the moisture and tears. When the eye is well washed by them they are carried off by other pipes, or tubes, into the nose, *b b*, and, running through a larger pipe, *c d*, end at *d*, which terminates at the nostrils, where they are evaporated, or dried up, and give us little further inconvenience.

69. Sometimes, in disease, this tube gets stopped, and then, unless the surgeon can unstop it, the tears run over the lower eyelid, on the cheek, and are very troublesome. To relieve this inconvenience a silver tube is sometimes let into the nose, to supply the place of the natural one, and the tears find their way through it.

70. That the eyelids may shut with greater exactness, and not fall into wrinkles when they are elevated or depressed, each ridge is stiffened with what is called a cartilaginous arch, that

is of a substance more thick and elastic than the skin, but not so hard as bone—just the kind of substance required for the purpose.

71. The eyelashes are like two palisadoes of short hair, and proceed from these cartilaginous edges, warning the eye of danger, protecting it from straggling dust, or motes, and warding off wandering insects.

72. The means provided for giving motion to the eye may 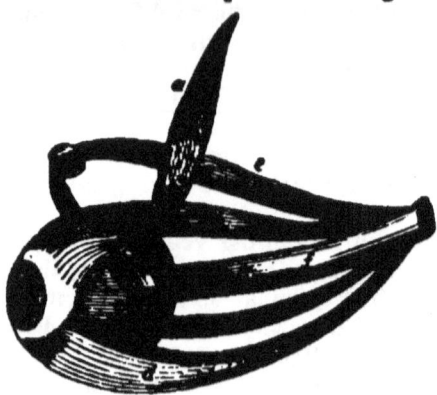 be seen in the figure, which is a side view of the muscles of the eye in their natural position; it shows, also, the position of the optic nerve, marked *f* The four straight muscles marked *a b c d*, enable us to raise and depress the eye, while the oblique muscle, *e*, which passes through a loop, gives us the power of moving it from side to side.

ANATOMY OF THE EYE.

73. The cut represents the section of an eye. The external 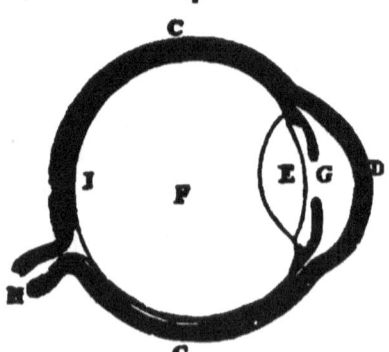 coat, which is represented by the outer circle, CCD, is called the *sclerotica*. The front part of this, D, is quite transparent, and is called the *cornea;* beyond this, towards C, above and below, it is white, called the white of the eye. The next coat, which is represented by the second circle, is called the *choroides;* this circle does not

ANATOMY OF THE EYE. 149

to round all the vacant space: G, is that which we call the pupil, and through this alone the light is allowed to enter the eye.

74. THE IRIS.—That part of the eye, which is of a beautiful blue in some persons, and in others brown, or almost black, is part of the *choroides*. It is composed of a sort of network,

 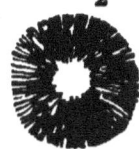

consisting of two different kinds of muscles, one *circular*, 1, and the other *radial*, 2. When a very luminous object is viewed the circular fibres contract, and the radial are relaxed, and thus the size of the pupil is diminished. On the other hand, when the objects are dark and obscure, the radial fibres of the iris contract, the circular are relaxed, and the pupil is enlarged, so that it admits a greater quantity of light. The cut represents the eye, after cutting away the sclerotic coat and cornea, to show the vessels and fibres of the choroid magnified. The whole of the choroid membrane is opaque, by which means no light can enter the eye but what passes through the pupil; but to render the chamber of the eye still darker, the part behind is covered over with a black substance, called the pigmentum nigrum (black paint).

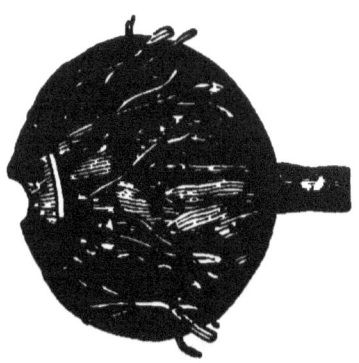

75. The third coat of the eye is called the *retina*, or network. It is a fine and delicate membrane, being an expansion of the optic nerve, which proceeds from the brain. It is spread like a network of exquisite delicacy, all over the concave surface of the choroides, at the back part of the eye; it terminates about half-way forward, and lies directly over the pigmentum nigrum.

76. The retina receives the images of objects which are depicted upon it by the rays of light that enter at the pupil. It is, of itself, transparent, and of an ash-coloured white, but appears black, on account of the dark coloured ligament behind it.

THE HUMOURS OF THE EYE.

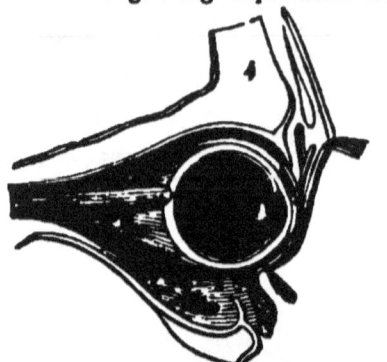

This engraving represents the section of an eye, as set in its bony orbit; 4 the crystalline lens, the aqueous and vitreous humours, the insertion of the optic nerve, and the attachment of the iris to the choroides, &c.

77. The coats of the eye, which invest and support each other, enclose three transparent bodies, called the *aqueous*, *crystaline*, and *vitreous* humours.

78. The humours of the eye are intended for refracting the rays of light, in the same manner as glass lenses. The vitreous humour fills all the space between the lens and the back of the eye; it is nearly of the substance of melted glass,—hence its name. The crystalline is in the shape of a double convex lens; and the aqueous, or watery humour, fills up that part of the eye between this lens and the cornea.

79. These lenses would not, however, perform their office, without the eye had the power of adjusting them to their proper focus, just as the lenses of a telescope are adjusted, by drawing out the sliding tube; for this purpose, membranes, in which some of the lenses are held, have the power of contracting or dilating occasionally, by which means they alter the shape of the natural lens, and shift it a little backwards, or a little forwards, so as to adapt its focal distance from the retina to the different distances of objects. Without this contrivance, or some equivalent arrangement, we should only see those objects distinctly that were at one distance from the eye.

80. It is impossible to say in what manner the image of any object painted on the retina of the eye, is calculated to convey to the mind an idea of that object; but to show that the

Images of the various objects which we see are painted on the retina, an artificial eye may be made by means of lenses, as in the case of a camera-obscura; or, we may take a bullock's eye, and, if from the back part we cut away the various coats, so as to leave the vitreous humour perfect, and then take a piece of white paper and hold the eye towards the window, the figure of the window will be drawn on the eye in an inverted position.

81. The paper, in this instance, represents the innermost coat, called the retina, and paper is used because it is easily seen through, whereas the retina is opaque;—transparency would be of no advantage to it. The retina, by means of the optic nerve, of which it is an expansion, conveys to the mind the views of every object that is painted on it.

82. MECHANISM OF VISION.—The rays of light proceeding from external objects are refracted in passing through the different humours of the eye, and converged to a point, or there would be no distinct picture drawn on the retina, and, of course, no distinct idea conveyed to the mind. As every point of an object A B C sends out rays in all directions, some rays, m n o,

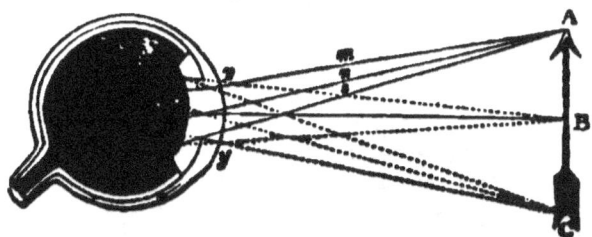

from each point on the side next the eye, will fall upon the cornea between x y; and, by passing through the humours of the eye, they will be converged, and brought to as many points on the retina, and will form on it a distinct inverted picture of the object.

83. All three of the humours have some effect in refracting the rays of light; but the crystalline is the most powerful, and that is a complete double convex lens. Its spherical aberration is corrected by increased density at its centre. The most perfect vision is from those rays which pass nearest the axis of the pupil.

84. It is not very easily explained, why we see things in their natural position, when the image of them forms an inverted picture upon the retina. It has been supposed that we acquire, by experience, the habit of seeing objects erect, but there are many striking facts to prove the contrary: persons who have been blind from infancy, and who have been suddenly restored to sight by a surgical operation, have not been led into the smallest mistake. How this correction is made, and how the mind perceives visible objects through the instrumentality of the eye is not known, the connection between the mind and the brain being screened from the view of man.

85. It is indeed wonderful that, in so small a space as the retina of the eye, the images of so many objects can be formed, and all with such perfect fidelity. The prospect from the top of St. Paul's, embracing, at least, a hundred square miles in extent, is reduced, by the arrangement of the various lenses of the eye, into a space not much larger than a pea. A stage coach, travelling at its ordinary rate for half an hour, when seen at a distance, passes in the eye over the twelfth part of an inch only, and yet the change of place is distinctly preserved through the whole progress.

86. The more we investigate the works of nature, the greater reason have we to admire the wisdom of its Author, and that wonderful adaptation of our organs in their minutest particulars to the general laws which pervade the universe. The mechanism of the eye affords a striking illustration of this remark. We have hitherto supposed the eye to be a lens, possessing the power only of enlarging and contracting its dimensions; and, from the previous description given of the rays of light, it may be considered as incapable of obviating the impression which must arise from their different modes of refrangibility; but here, the use of the wonderful structure of parts, and the different fluids of the eye is clearly seen. The eye is in fact a compound lens, *each fluid has its proper re*frangible power. The *shape* of the lenses is altered at will, according to the distance of the object, and the three substances having the proper powers of refrangibility, the effects of an *achromatic* telescope, are without difficulty, produced in the eye.

87. The CAMERA OBSCURA consists of a convex lens, placed in such a position that all the light from without shall pass through it into a darkened chamber, where objects will be represented in their natural colours, either upon the wall or upon the white surface of a table placed to receive them.

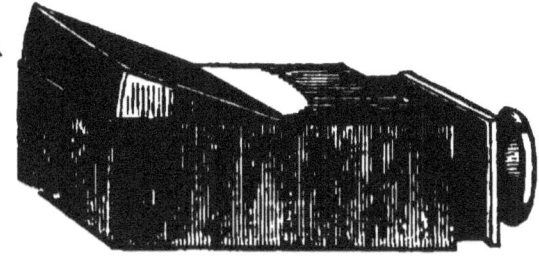

88. A portable camera obscura may be made with a small box having the form of a parallelopipedon, in one end of which a tube must be fitted, containing a lens, and made to slide backwards and forwards, so as to suit the focus. Within the box is a plane mirror reclining backwards from the tube in an angle of 45 degrees; the top of the box is a square of unpolished glass, upon which, from beneath, the picture will be thrown, and may be seen by raising the lid A.

89. The image of the objects would naturally be formed at the back of the box opposite to the lens; but to prevent this, and to throw them on the top, the mirror must be so placed that the angle of incidence shall be equal to the angle of reflection. In the box, according to its original make, the top is at right angles to the end, that is, at an angle of 90 degrees; therefore, the mirror is put at half 90 deg. that is, 45 deg. Now the incident rays falling upon a surface which declines to an angle of 45 degrees, will be reflected at an equal angle of 45 degrees, the angle which the glass top of the box bears with respect to the mirror.

90. The form of the camera obscura, used as a public exhibition, is as follows: F D is a large wooden box stained black on the inside, and capable of containing from one to eight persons; C B D F is a sliding piece having a sloping mirror C D, and a double convex lens F D, which may, with the mirror C D, be slid up or down so as to accommodate the lens to near and distant objects. When the rays proceeding from an object without, as A B, fall upon the mirror, they are

154   THE CAMERA OBSCURA.

reflected upon the lens F D, and brought to foci on the bot-

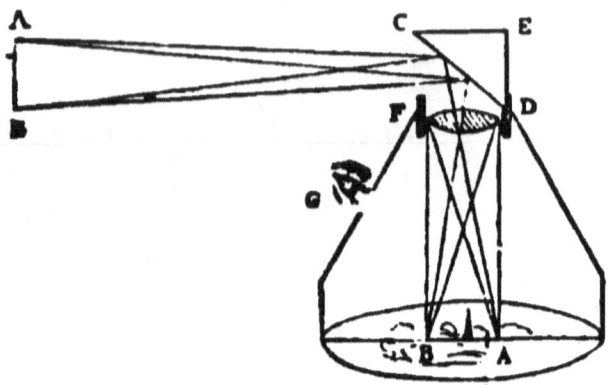

tom of the box, or upon a table placed horizontally to receive them, which may be seen by the spectator whose eye is at E.

91. The CAMERA LUCIDA consists of a glass prism C D E F having four sides inclined. The side C D being exposed to the object to be delineated, rays pass through it and fall upon the sloping side D E; from this they are reflected to the side F, and finally pass out of the prism to the eye at F; now

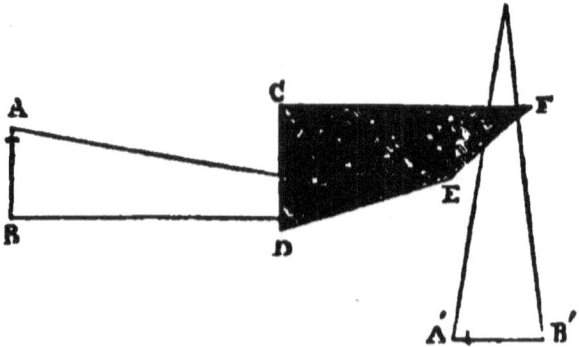

from the direction in which the rays enter the eye, it receives them as if coming from an image at A' B', and if a sheet of paper be placed below the instrument, a perfect delineation of the object may be traced with a pencil.

92. The *magic-lantern* consists of a sort of tin box, within which is a lamp or candle; the light of this passes through a great plano-convex lens, C D, strongly reflected by the reflector I, placed in a tube fixed in the front. This strongly illuminates the objects which are painted on slips of glass, and placed

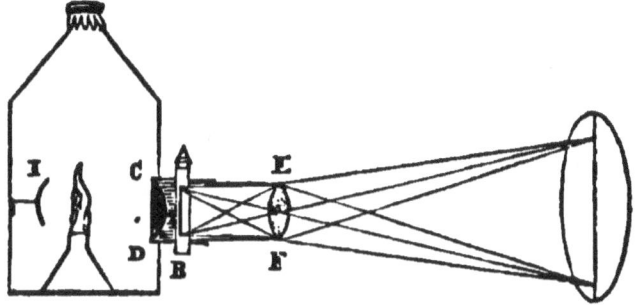

before the lens in an inverted position, and the rays passing through them and the lens E F, fall on a sheet, or other white surface, placed to receive the images. The glasses on which the figures are drawn are inverted, in order that the images of them may be erect.

93. Between the phantasmagoria and the magic-lantern there is this difference: in common magic-lanterns, the figures are painted on transparent glass, consequently the image on the screen is a circle of light having a figure or figures on it ; but in the phantasmagoria, all the glass is made opaque, except the figure, which, being painted in transparent colours, the light shines through it, and no light can come upon the screen but what passes through the figure. There is no sheet to receive the picture; but the representation is thrown on a thin screen of silk placed between the spectators and the lantern. The images are made to appear approaching and receding by removing the lantern farther from the screen, or bringing it nearer to it.*

* Read Illustration of Optics.

ASTRONOMY.

CHAPTER XXXIV.

HISTORY OF ASTRONOMY—PYTHAGOREAN, PTOLEMAIC, CO-
PERNICAN, AND TYCHONIC SYSTEMS.

1. ASTRONOMY is the science which demonstrates to us the laws by which the heavenly bodies are governed, explains their motions, periods, and magnitudes, and is applied to Geography, Chronology, and Navigation.

2. It is divided into Pure and Physical.—Pure or Plane Astronomy treats of the planetary motions, without any reference to the cause of such phenomena.—Physical Astronomy

investigates the causes of the motions, periods, eclipses, &c., of the heavenly bodies.

SKETCH OF HISTORY.

3. Astronomy is one of the most ancient of the sciences. Josephus ascribes to Seth and his posterity an extensive knowledge of the science. In profane history it may be traced to Chaldea or Egypt, and the Chinese have traditionary accounts of its cultivation. The Egyptians had their monument of Osymandias, on which we are told was a golden circle, 60 cubits in circumference, divided into 365 parts, according to the days of the year.

4. Astronomy was co-eval with the earliest history of Greece; several of the constellations are mentioned by Hesiod and Homer. Thales the Milesian greatly improved the science, and predicted an eclipse; Anaxagoras also taught that the moon consisted of hills and valleys, and water, while to Pythagoras we are indebted for the true doctrine concerning the universe and motions of the planets.

5. In the second century of the Christian era, Ptolemy invented a system which was implicitly followed by all nations, till within 300 years of the present time. This system supposed the earth to be in the centre, with the sun and planets revolving round it, in the following order—Moon, Mercury, Venus, Sun, Mars, Saturn, Jupiter.

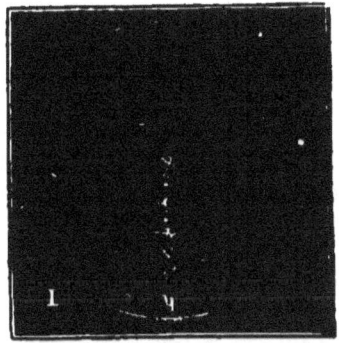

6. From this time Astronomy slept till 800 years after Christ, when among the Arabians the Caliph Al Mansur introduced a taste for the science into his empire. His grandson Al Manum made many observations, and Ebn Younis observed three eclipses; while among the Moors of Spain, Arzachel and Alhazen advanced the theory of refraction. Ulug Beg, grandson of Tamerlane the Tartar prince, is reported to have constructed

instruments of a very large size for the advancement of the science.

7. From the year 1471 to 1500 John Müller and John Werner published many treatises on the Mathematics, and about the same time Copernicus appeared, who conceived doubts concerning the Ptolemaic system, and soon demonstrated the system of Pythagoras to be the true system of the universe. This is now called the SOLAR SYSTEM.

8. Tycho Brahe attempted to improve upon Copernicus, in his design to reconcile some passages of Scripture with his doctrine, but failed. The earth was placed in the centre of the orbits of the moon and the sun, whilst the sun was the centre of the orbits of the rest of the planets; the whole revolving round the earth in 24 hours. Kepler came next, and ascertained for a certainty that all the planets move in elliptical orbits, that the squares of the times of their revolutions were to one another as the cubes of the distances, and that an imaginary line, connecting the planet with the sun, sweeps over equal areas in equal times.

Tychonic System.

9. In the beginning of the 17th century, telescopes were invented, and being improved by Galileo enabled him to discover the inequalities on the moon's surface, Jupiter's satellites, the ring of Saturn, and the spots on the surface of the sun, by means of which he discovered the revolution of that luminary round its axis.

10. In the year 1638 Helvetius, who flourished at Dantzic, furnished a more complete catalogue of stars, than any that had before appeared. Huygens and Cassini discovered the satellites of Saturn; and Newton demonstrated from physical considerations the motions by which the heavenly bodies are regulated. Lately Sir John Herschel has made important observations on the celestial phenomena of the southern hemisphere.

THE SOLAR SYSTEM.

CHAPTER XXXV.

THE SUN, MERCURY, VENUS, ETC.

11. THE SOLAR SYSTEM, or system of the sun, consists of sixteen planets and their satellites moving round the sun in orbits *slightly elliptical*. This system has been indifferently called the Pythagorean, the Copernican and Solar system: Pythagoras being its originator, and Copernicus and Newton its demonstraters in later times.

12. ORDER OF THE PLANETS.—The planets move in their orbits from west to east, in the following order:—1, MERCURY;

2, Venus; 3, Earth; 4, Mars; 5, Jupiter; 6, Saturn; 7, Herschel; 8, Neptune, and the eight asteroids between the orbits of Mars and Jupiter are *Juno, Pallas, Ceres, Vesta, Astræa, Hebe, Iris, and Flora.* This solar system also includes the *comets,* in very narrow elliptical orbits. It assumes each planet to turn on its *axis,* and to carry its satellites round with it, as it moves in its *orbit.*

13. Planetary Forces.—The power which causes the planets to move round the sun is of two kinds, and is called the *centripetal* force, and the *centrifugal* force. The first is the power which the sun has of drawing bodies to him, and is called *attraction;* the other is the force which the planet has impressed upon it to move onwards in a right line,— which two forces, acting in opposition, cause it to move in a circle. (See *Mechanics,* p. 23).

14. Shape of the Planetary Bodies.—Each planet is more or less flattened at its poles. This arises from the rapidity with which it turns on its axis, causing it to swell out in the same way as a hoop would bulge in the middle, if twisted into motion.—(See *Mechanics.*)

15. The Orbits of the Planets.—The orbits of the different planets do not all lie in what is termed the plane of the ecliptic, as they appear to do in orreries, and in the representations generally given of the solar system. If we suppose a plane to pass through the earth's orbit, and to extend in every direction, it would trace a line in the starry heavens which is called the ecliptic, and the plane itself is called the *plane of the ecliptic.* The orbits of all the other planets do not lie in this

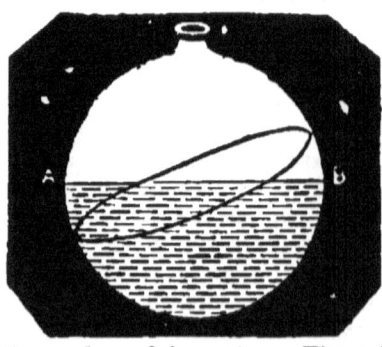

line, one half of each orbit lying above it, and the other half falls below it, as in the annexed cut. The cut represents a globe filled with water, the surface of which, A B, may represent the plane of the ecliptic, the line seen within the vessel the orbit of a planet, inclined, and not parallel to the line which represents the surface of the water. The points of intersection where the lines cut each otner are called the nodes. That point in which the orbit cuts the ecliptic going northward is the ascending node, southward the descending node. The ECLIPTIC is supposed to be divided into 360 degrees, or 12 signs, called the signs of the zodiac, which have received the following names:—Aries, or the Ram ♈ ; Taurus, or the Bull ♉ ; Gemini, or the Twins ♊ ; Cancer, or the Crab ♋ ; Leo, or the Lion ♌ ; Virgo, or the Virgin ♍; Libra, or the Scales ♎; Scorpio, or the Scorpion ♏ ; Sagittarius, or the Archer ♐ ; Capricornus, or the Goat ♑ ; Aquarius, or the Water-bearer ♒ ; Pisces, or the Fishes ♓.

16. The figures A D B E represent the form of a planetary orbit. The longer diameter is A B, the shorter D E. The two points, F and G, are called the *foci of the ellipse*, around which two central points the ellipse is formed. The sun is not placed at C, the centre of the orbit, but at F, one of the foci of the ellipse; when the planet, therefore, is at A, it is  nearest the sun, and is said to be in its *perihelion;* its distance from the sun gradually increases till it reaches the opposite point, B, when it is at the greatest distance from the sun, and is said to be in its *aphelion;* when it arrives at the points D and E of its orbit, it is said to be at its *mean distance.*

17. The line A B which joins the perihelion and aphelion, is called the line of the *apsides,* and also the greater axis or the *transverse* axis of the orbit. D E is the *lower* or *conjugate* axis; F D the mean distance of the planet from the sun; F C or G C the *eccentricity* of his orbit, or the distance of the sun from the centre. F is the *lower focus,* or that in which the sun is placed; G the *higher focus;* A the *lower apsis;* and B the *higher apsis.*

THE SUN, ☉.

18. THE SUN is a spherical body, whose diameter is 882,000 miles, and is the centre of the planetary system. He is the source of light and heat, and revolves round his axis in 25

days, 10 hours. He is 95,000,000 miles from the earth, and 1,384,472 times larger. His surface contains 2,432,800,000,000 square miles; and although his density is little more than that of water, he would weigh 329,000 globes such as our earth. The sun is not a globe of fire, as was formerly supposed, but a solid body, surrounded with an atmosphere from which our light and heat emanate. It is assumed by some, that heat is generated by this luminous atmosphere being projected in atoms from the sun, and coming in contact with the atmosphere of the planets, and that light is the same particles.

19. MOTIONS OF THE SUN.—The sun has two *real* and two *apparent* motions. The *first* real motion of the sun is a small circular one round the *centre of the orbits of all the planets*, which motion is supposed to be occasioned by the various attractions of these surrounding planets. The *second* real motion is that round his own *axis*.

20. The sun's *first apparent* motion is the diurnal motion from east to west, and arises from the motion of the earth upon its axis. The *second* is the sun's apparent *annual* motion in the ecliptic, which arises from the *earth's motion in her orbit*. A spectator in the sun would see the earth move from west to east, as we see the sun move from east to west.

21. SPOTS ON THE SUN.—Astronomers, by the aid of telescopes, have discerned spots on the sun, moving from west to east and exhibiting numerous changes from day to day. The magnitude of these spots is very great, sometimes measuring a twentieth part of the sun's diameter, or 45,000 miles. They are supposed to be openings in the atmosphere, through which the real body of the sun may be discerned; thus the sun appears to be a very large and glorious planet, in all probability inhabited, like the rest of the planets, by beings whose organs are adapted to the peculiar circumstances of their situation.

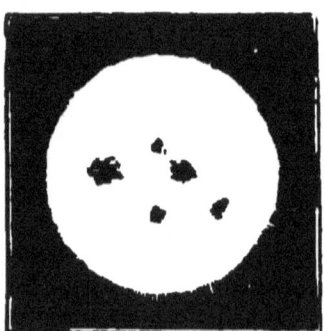

CHAPTER XXXVI.

MERCURY, ☿.

22. Mercury is the first planet in order next the sun, his distance being nearly 37,000,000 miles. He moves in his orbit at the rate of 109,000 miles an hour, and goes round the sun in 87 days, 23 hours, 15 minutes, and 44 seconds, which space of time is the length of his year.

23. Mercury appears small to the eye, but emits a very bright white light, and keeping always near the sun, is never seen longer than an hour and 50 minutes before the *rising* or *after* the setting sun.

Mercury

24. This planet is not much larger than the moon, and when viewed through a telescope exhibits similar *phases* or changes, being sometimes *horned* and sometimes *gibbous*, but never entirely full, because its enlightened side is never turned directly towards us. These phases prove, that this planet is a dark opaque body, and does not shine by its own light.

25. When Mercury is directly between the earth and the sun, he appears on the sun's disc as a black spot. His passing before the body of the sun is called a transit, the last two of which occurred May 5, 1832, and November 7, 1835.

26. The laws of the movements of Mercury are exceedingly complicated; he does not move in the plane of the ecliptic, deviating at times more than 5 degrees. His motion round the sun is swifter than that of any other planet yet discovered, being at the rate of 109,000 miles an hour, which is more than 30 miles a second. The density of Mercury is also greater than that of any other planet, being 9 times that of water, or equal to that of lead. We can discover no spots on the face of Mercury, consequently, have no means of computing the time of

his diurnal motion very accurately. He is assumed, however, to turn on his axis in 24 hours, 5 minutes, 28 seconds. No satellites or moons have yet been discovered to attend him.

OF VENUS, ♀.

27. Venus is one of the most beautiful stars in the heavens, exceeding all the planets in splendour, and casting a sensible shadow. She presents to us, when viewed through a telescope, the same phases as the planet Mercury; her surface exhibits spots like the moon, and she is thus discovered to revolve on her axis in 23 hours, 21 minutes.

28. The axis of Venus is considerably inclined to her orbit, and thus the variation of her seasons must be great, and the length of her days and nights differ very much in proportion to each other.

29. The diameter of Venus is 7,700 miles; her distance from the sun, 68,000,000 miles; she performs her orbit in 224 days, 16 hours, 49 minutes, which is the length of her year; she moves at the rate of 80,060 miles an hour.

30. Twice in the course of 120 years Venus passes over the disc of the sun, and in the year 1761 Captain Cook, aided by several astronomers, went to the South Seas to observe the transit. The next transit will be in 1882.

31. Venus is a morning star when she appears west of the sun, and thus rises before him. In poetical language she is called Lucifer or Phosphorus; and when she appears east of the sun, and sets after him, she is called Hesperus or Vesper. She is 290 days an evening star, and a morning star somewhat longer.

32. M. Schroter discovered several mountains on Venus, and found that, like those of the moon, they were always highest in the southern hemisphere. They are supposed to be of considerable altitude. Her atmosphere is of considerable density; and is supposed to rise to a height of 16,000 feet.

33. A body which weighs 1 pound at the Equator of our earth, would, if removed to that of Venus, weigh only 0.98 of a pound. Her proportion of light and heat is about 1.91 greater.

CHAPTER XXXVII.

OF THE EARTH AND MOON.

34. THE EARTH on which we live is the third planet from the sun. Her polar diameter is 7,889 miles; her equatorial diameter 7,926 ; and her mean diameter is 7,912 miles, and her circumference 24,872 miles; her distance from the sun is 95,000,000 miles.

35. SHAPE OF THE EARTH. — The earth is of a spherical shape, which is proved—first, because her shadow, when thrown upon the moon at the time of an eclipse, is always circular; second, when a ship leaves the shore, the hull becomes first invisible at a distance; thirdly, in sailing towards the north, from the equator, the pole star becomes more and more elevated, which clearly shows a gradual rotundity of the earth's surface. The earth is not, however, a perfect globe, her polar diameter being about 27 miles less than her equatorial. Hence she is called an *oblate spheroid*.

36. THE EARTH'S MOTIONS.—The earth has two motions: first, her *annual* motion, which is the motion of her whole body round the sun, performed in 365 days, 5 hours, 48 minutes, and 51 seconds; second, her *diurnal* motion, which she completes in 24 hours, the length of her day. By the annual motion is produced the seasons ; by her diurnal is caused day and night. The side of the earth turned to the sun, and receiving its light, has the day ; the side which is turned from him has night. There is also a third motion, called the precession of the equinoxes, which is a slow motion of the two points, where the equator cuts the ecliptic, which is found to move backwards about 50 seconds in a year.

37. INCLINATION OF THE EARTH'S AXIS.—The axis of the earth is inclined to the plane of her orbit 66½ degrees. It

166 INCLINATION OF THE EARTH'S AXIS.

is always parallel to itself, and directed constantly to the same point in the heavens. This, combined with the earth's motion in her orbit, gives rise to the phenomena of the SEASONS, and the varied lengths of days and nights. In the spring equinox,

the circle which separates the light from the dark side of the globe passes through the poles, as at Mar. 21, and Sep. 21, when days and nights are equal. During summer in the northern hemisphere, it passes 23½ deg. beyond the pole, as on June 21, when the days are 16 hours long in the latitude of London, while the sun never sets within the circle 23½ degrees from the pole.

38. At the autumnal equinox, the days and nights are again equal, the earth being in the position Sep. 21, half of each parallel has the light of the sun. In winter, the North Pole is turned away from the sun, Dec. 21, whilst the South Pole is enlightened; consequently, we have six months' darkness at the northern pole, and very short days in the northern latitudes; while the South Pole enjoys six months' sunshine, and southern latitudes experience summer.

CHAPTER XXXVIII

THE MOON

39. The Moon is an opaque body, nearly spherical, which receives its light from the sun. Her diameter is 2,165 miles, and her circumference 6,851; she is 240,000 miles from the earth, and performs her rotation on her axis in 29 days, 12 hours, 44 minutes, and 3 seconds. She is the satellite or attendant on our earth, and supplies us with light during the night. The time of the moon's revolution round her axis and our earth, being constantly the same, she has always the same side turned towards us; she moves, in her orbit, 2,275 miles an hour; performing near 13 revolutions round our planet in the year.

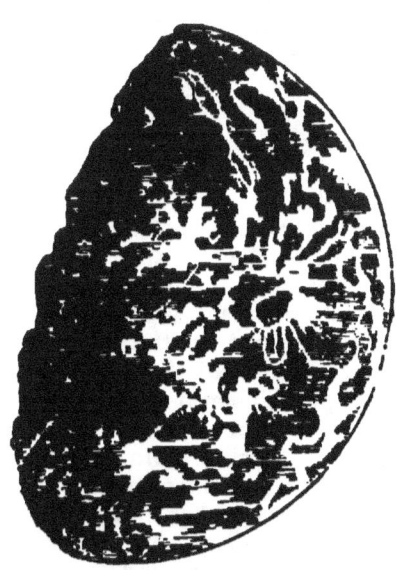

40. PHASES OF THE MOON.—The phases of the moon, as they appear at eight different points of her orbit, are represented in the accompanying figure, where *s* represents the sun, the earth being in the centre, and *a b c*, &c., the moon's orbit. When the moon is at *k*, in conjunction with the sun, her dark side being entirely towards the earth, she will be invisible, as at *a*, and is then called the new moon. When she comes to her first octant at *i*, a quarter of her enlightened hemisphere will be turned towards the earth, and she will then appear horned, as at *h*. When she has run through the quarter of her

orbit, and arrived at *q*, she shows us the half of her enlightened hemisphere, as at *g*, when it is said she is one half full. At *p*

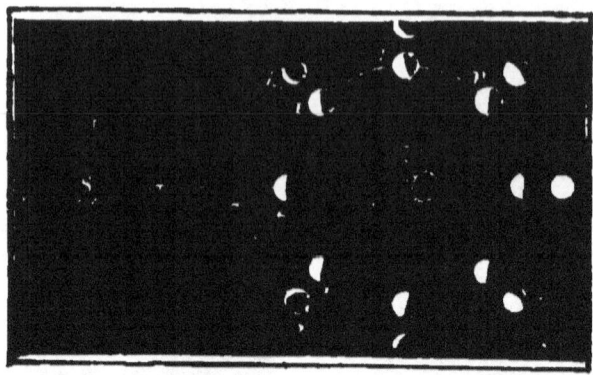

she is in her second octant, and by showing us more of her enlightened hemisphere than at *g*, she appears gibbous. At her opposition at *o*, her whole enlightened side is turned towards the earth, when she appears as at *e*, and she is said to be full; having increased all the way round. On the other side she decreases again all the way from *e* to *a*; thus, in her third octant, part of her dark side being turned towards the earth, she again appears gibbous. At *m* she appears still farther decreased, showing again exactly one-half of her illuminated side. But when she comes to her fourth octant, she presents only a quarter of her enlightened hemisphere, and again appears horned. And at *a*, having now completed her course, she again disappears, and becomes a new moon again, as at first.

41'. ORBIT OF THE MOON.—The orbit of the moon is in the form of an ellipse, not in the same plane with the ecliptic, but inclined to it, in an angle of 5 degrees 9 min. These two planes, therefore, cut each other at two points, which points are called the nodes. These intersections are not always at the same place, but at a point further westward every lunation, which, repeated for 18 years and 228 days, brings her to cross the ecliptic again in the same point. It is only when a new or full moon happens at or near these nodes, that an eclipse of the sun or moon can take place. When the moon is at her greatest distance from the earth, she is said to be in her *apogee*; when at her least, in her *perigee*

42. TELESCOPIC APPEARANCE.—The face of the moon, when viewed through a telescope, appears diversified with irregularities. These are supposed to be hills and valleys, and as these spots reflect a triangular shadow, in a direction from the sun, they are known to be mountains, and their heights can be calculated from the length of their shadows; while, on the contrary, the cavities are always dark on the side next the sun, and illuminated in a contrary direction. Volcanoes in eruption have also been clearly distinguished in the moon by Dr. Herschel and others. One showed an actual eruption of luminous matter, which continued visible for some time.

43. LIBRATION IN LONGITUDE.—Sometimes one portion of the moon's face or disc on the eastern side is carried entirely out of sight, when a similar part of the western side is brought forward. In another part of her revolution the contrary will be seen; the portion so brought forward on the western side will disappear, and the eastern portion be brought into view. This irregularity is called the moon's libration in longitude.

44. LIBRATION IN LATITUDE.—Another sort of libration, called the libration in latitude, arises from the axis of the moon being inclined to the plane of her orbit; for sometimes in like manner we see more or less of her polar regions, in different parts of her revolution, as her poles are inclined more or less towards the earth.

It was formerly supposed that the moon had no atmosphere; but certain astronomers have supposed that she has one, of about a third the density of ours.

45. HARVEST MOON.—The moon, when full in harvest, rises several nights sooner after sunset, and with less difference of time than at any other full moon during the year. The reason of this is, because she is then in Pisces and Aries, and these signs rise in a shorter time than others. In winter she is also in these signs, when she rises about noon; in spring, when she rises with the sun; and in summer, when she rises at midnight.

L

CHAPTER XXXIX.

MARS AND HIS PERIODS—PALLAS—CERES—JUNO, AND VESTA.

46. The next planet in order to the earth is Mars, distinguished in the heavens by the redness of his colour, which is attributed to the density of his atmosphere. His figure is that of an oblate spheroid, like that of the earth. His orbit is considerably eccentric, namely $\frac{1}{11}$ of his diameter, which is more than eight times the eccentricity of the orbit of the earth. Thus he is very much nearer the earth at one time than another.

47. The diameter of Mars is 4,189 miles; his distance from the sun 144,000,000 miles. He moves round the sun in 686 days, 23 hours, and 30½ minutes, and turns on his axis in 24 hours, 39 minutes, and 21 seconds. The inclination of the orbit of Mars to that of the earth is 1° 51′ 6″.

48. TELESCOPIC APPEARANCE OF MARS.—Mars presents many spots and irregularities on his surface, which change their form. He has also an increase and decrease, like Venus and Mercury, but is never horned like them. He presents a remarkable brightness about his North and South Pole, supposed to arise from the snow and ice accumulated in those parts, as they are observed to grow less when turning from the sun, and largest when turning out of their 12 winter months. No moon or secondary planet has been discovered about Mars, but there is no proof he does not possess one, for its extreme minuteness might prevent its being seen.

49. THE ASTEROIDS.—From the great space which subsists between the orbits of Mars and Jupiter, it was long thought that some planetary bodies were situated therein. Their discovery by the aid of the telescope verified the supposition.

50. CERES, when viewed through a telescope, appears of a

ruddy colour, and looks like a star of the eighth magnitude. She was discovered by M. Piazza, in 1801, and goes round the sun in 4 years, 221 days, 32 hours, and 9 seconds.

51. PALLAS was discovered in 1802, by Dr. Olbers, of Bremen. Her diameter is only 80 miles according to Herschel, but Schroeter thinks it to be 2,099. She is 262,000,000 miles from the sun, which she goes round in 4 years, 7 months, and 11 days.

52. The orbits of some of the asteroids intersect each other, namely those of Ceres and Pallas, as in the diagram; Pallas coming nearer to the sun at her *perihelion* than Ceres, and going to a greater distance at her *aphelion*. This is a very singular and unaccountable circumstance in regard to planetary orbits.

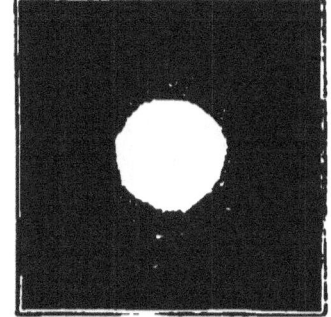

53. JUNO was discovered by Mr. Harding, at Lilenthal in Germany: her mean distance from the sun is 254,000,000 miles. Her annual revolution 4 years and 128 days. Her diameter, according to Schroeter, 1,428 miles.

54. VESTA was discovered March 9, 1807, by Dr. Olbers, situated between the orbit of Juno and Mars. The year is about the length of 3 years and 66 days of our time, and diameter about 1,800 miles, but neither have been accurately ascertained.

55. Another peculiarity with regard to these planets is, that they revolve nearly at the *same mean distance from the sun*, that of Juno being only 254,000,000 and that of Pallas 262,000,000 miles. This is a very different arrangement to that of the other planets, whose mean distances are immensely different from each other. Other asteroids have been discovered since 1845, the names and dates of which are as follows:—Astræa, in 1845; Hebe, Iris, and Flora, in 1847; Metis, 1848; Hygeia, 1849; Parthenope, Victoria, and Egeria, 1850; Irene, Eunomia, 1851; Melpomene; Massilia, Fortuna, Lutetia, Thetis, Thalia, Calliope, Psyche, 1852, Euterpe, Phocea, Proserpine, Themis, 1853; Urania, Amphitrite, Pomona, Bellona, Polyhymnia, Euphrosine, 1854; Fides, Circe, Atalanta, Leucothea, 1855; Leda, Lætitia, Harmonia, Daphne, Isis, 1856; and others, now making upwards of 60, in 1862.

OF JUPITER, ♃, AND HIS SATELLITES.

56. Jupiter is next to Venus in brightness and splendour of appearance, and is the largest of all the planets. He is 89,000 miles in diameter; his distance from the sun is 490,000,000 miles; he turns upon his axis in 9 hours, 55 minutes, and 49½ seconds, being at the rate of 29,000 miles an hour.

57. From this planet turning so swiftly on its axis, its figure is much more oblate than that of any planet in the system, he being more than 6,000 miles at his equatorial than at his polar diameter. The length of his year is 4,332 days.

58. THE AXIS OF JUPITER is so very perpendicular to his orbit, that he has no difference of seasons, and this is wisely ordained; for, were his axis much inclined, vast tracts round the poles would be deprived of the sun's influence for six years together. In his polar regions there is perpetual winter, and in his equatorial perpetual summer.

59. BELTS OF JUPITER.—When viewed through a telescope, Jupiter exhibits in his body a series of bright belts, horizontal and parallel, from one to eight in number. Their breadth is not always the same: sometimes one is seen to widen considerably, and another to become narrower. They are supposed to be the clouds round the planet, collected about his equatorial parts by the rapidity of his diurnal motion. These belts were first discovered by Fontana; they were afterwards more particularly observed and delineated by Cassini.

60. In addition to these belts, Hooke discovered spots of various sizes and shape. The largest, which was observed by Cassini in 1665, appeared round, and moved with the greatest velocity when near the middle of the planet, but slower when it reached the circumference, similar to the spots on the sun. This spot determined the revolution of the planet on its axis

61. SATELLITES OF JUPITER.—Jupiter is attended by four satellites or moons, revolving about him in different times. The first in 1 day 18 hours, the second in 3 days 13 hours, the third in 7 days 14 hours, the fourth in 16 days 16 hours. They were first seen by Galileo, in the year 1610. These satellites are continually suffering eclipses, by passing behind the body of the planet; and as these occultations are discoverable from the earth, and are seen at the same instant of time from whatever part of the world the observer may be placed, they are made use of to ascertain the longitude.

62. To find the longitude by the immersions of Jupiter's satellites, it is necessary to look in the English Ephemeris, where all the eclipses of Jupiter's first satellite, both immersions and emersions, are calculated according to the meridian of London. If then by a good telescope and correct chronometer, we attend to the immersions of the satellite; according to our distance east and west from the meridian of London, so will the hour of its appearance differ from that of the calculation. Now as 15 degrees of motion under the equator answers to one hour of time, so by the difference of time we measure the difference of distance, which gives us the longitude; or, in other words, tell us exactly how far we are east or west from the meridian of London.

63. Sometimes the satellites pass *before* the body of the planet; this is called a *transit*, when *behind* an *occultation*. It was from observing the eclipses of Jupiter that light was discovered not to be propagated instantaneously, though it moved with extreme velocity.—(See *Optics*, page 125).

64. DENSITY OF JUPITER.—The density of the planet, compared with that of water, is as $1\frac{1}{5}$ to 1; that is, it is a small fractional part denser than water. Its mass, compared with that of the sun, is as 1 to 1,067; compared with the earth, as 312 to 1; that is, Jupiter could weigh 312 globes of the same size and density as the earth. The plane of his orbit is inclined to the plane of the ecliptic, at an angle of 1° 18′ 51.″

SATURN.

65. SATURN is the sixth planet in the order of our system, and is not so bright as Jupiter, but when viewed through a

good telescope makes a more remarkable appearance than any of the planets, on account of a luminous ring which surrounds him.

66. This ring was first discovered by Galileo, and appeared like two small globes, joined one on each side to a very large one; having viewed him for two years he was surprised to find him quite round like any other planet, but in the course of time he resumed his original appearance.

67. About 40 years after Galileo, Huygens, by the aid of a telescope magnifying 100 times, discovered that this appearance of the planet was owing to a broad thick ring, or rather two rings one within the other, which some supposed to be an assemblage of planets, and others a permanent bright cloud.

68. These rings are detached from the body of Saturn, in such a manner, that the distance between the innermost part of the ring, and its body, is equal to the breadth; both the outward and inward rim of which is projected into an ellipsis, more or less oblong according to the different degrees of obliquity with which it is viewed. The reason of the disappearance of the ring to Galileo was because his eye had got upon the plane of it, and it became invisible. The breadth of the inner ring is about 20,000 miles.

69. Dr. Herschel has demonstrated that these rings have a rotation on their axis, and revolve round his body in about 10 hours; and he has observed that they throw a strong shadow on the body of the planet, inferring from thence, that the substance of the ring is no less solid than that of the

planet. Later astronomers have supposed these rings to transmit the sun's light.

70. Saturn has zones or belts similar to those of Jupiter, and originating from the same cause, viz., clouds floating in his atmosphere, which is observed to be very dense; he has also seven satellites or moons, which move round him nearly on the plane of the rings, and in certain positions appear like excrescences or roughnesses there.

71. The diameter of Saturn is by some said to be 77,680, by others 79,000 miles, or about ten times the diameter of the earth: he is near 30 years in performing his annual revolution, and 900,000,000 miles from the sun.

URANUS OR HERSCHEL.

72. This planet was discovered by Dr. Herschel in 1781. It has six satellites. It was named by Dr. Herschel "Georgium Sidus," in compliment to George II.

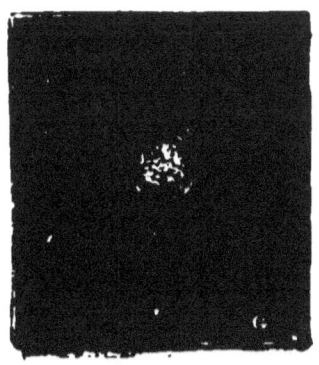

73. Its distance from the sun is 1,900,000,000 miles; its diameter is 35,170 miles, and its periodical revolution nearly 84 years. Its orbit is inclined to the plane of the ecliptic, 46 minutes 29 seconds. It is 82 times the bulk of our earth.

NEPTUNE.

74. In the year 1846 this planet, making the eighth and most distant planet of our solar system, was discovered by Mr. Galle, of Berlin. It had been predicted by Mr. Adams in England, and by Le Verrier in France.

75. The diameter of Neptune is 37,000 miles; its distance from the Sun 2,800,000,000 miles, being 900,000 miles more than Uranus; and its year 166 of ours.

CHAPTER XI.

OF THE TIDES.

76. Kepler first shewed this phenomenon of nature to be owing to the moon's influence upon the waters; and Newton demonstrated it to be a necessary result of gravitation.

77. It is easy to suppose, that although this attraction cannot alter the shapes of the solid parts of the globe, yet it may nevertheless produce certain effects upon the fluid parts. Thus then it is, the ocean is drawn towards the moon, and it is therefore high water at the place perpendicularly under the moon, or about 3 hours after the moon crosses the meridian.

78. If we fix a string to the side of a flexible hoop, and thereby swing it round in a circle, we readily conceive how the part next the hand would bulge out or swell, by the drawing of the string, and how the opposite would fly off or swell by the centrifugal force, it being least drawn in, and how the intervening parts of the hoop would become depressed and flattened.

79. Thus it is with the ocean; that part of it which is im-

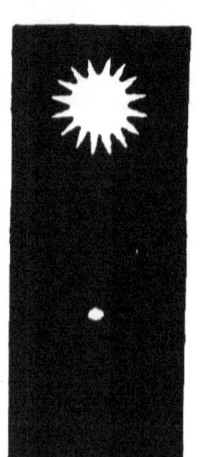

mediately under the moon is raised by its attraction up into a swell, and that part which lies on the opposite side of the earth, is thrown up into a similar swell by the motion of the earth in its orbit, by the centrifugal force. (See Diagram G.)

80. The sun has also some action on the waters, but only in the proportion of three to ten, in consequence of his great distance from the earth, compared with that of the moon.

81. Sometimes it happens, at the full and change of the moon, that the sun's attractive force is united with that of the moon, and increases it (See Diagram G). This union is productive of what are called SPRING TIDES. At other times, during the

half moons, the attractive power of the sun and moon counteract each other, and produce what are called NEAP TIDES (See Diagram F.)

82. The height of the tides at different places depends in a great measure upon the local situation of such places, and the varying depth of the sea. At the poles of the earth, the diurnal tides do not exist, for the moon never attains any considerable elevation above the horizon in these climates, and therefore the tides are scarcely perceptible.

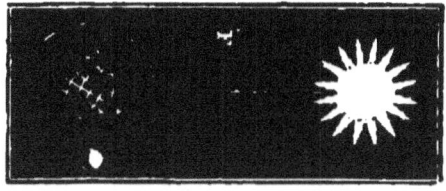

ECLIPSES.

83. There are two kinds of eclipses; those of the sun and those of the moon. An eclipse of the sun is called a solar eclipse, and is occasioned by the moon coming between the earth and the sun, and obstructing our view of it (see diagram M). A lunar eclipse is occasioned by the dark body of the earth coming between the sun and the moon, and thereby depriving the latter of the sun's light, in consequence of which she appears to us as a dark body (see diagram L).

84. Eclipses are either *total* or *partial*, or what is called *annular*. A *total* eclipse is when the whole of the body of either of the sun or moon is in darkness. A *partial* eclipse is when only a part of the sun or moon appears dark. An *annular* eclipse (so called from annulus, a ring) is when a portion

of the sun is visible quite round the moon as a luminous ring This kind of eclipse can only happen when the moon is neai her apogee, or greatest distance from the earth, when she is seen under a smaller angle than the sun.

85. Total eclipses of the sun only happen when the moon is in perigree, or nearest to the earth; and when a partial eclipse takes place when the centre of the moon does not pass apparently over the centre of the sun. An eclipse of the sun can only happen at the time of new moon, that is, when the moon, the sun, and the earth are in one straight line, or in conjunction; but an eclipse of the sun does not happen at *every* conjunction, because of the obliquity of the moon's orbit to the plane of the ecliptic, and only when she is near or at one of her *nodes*.

86. As the earth is an opaque body, enlightened only by the sun, it must cast a shadow towards that side which is furthest from the sun. If the sun and earth were of the same size, this shadow would be cylindrical, and would extend to an infinite distance; but as the sun is much larger than the earth, the shadow of the latter must be conical, ending in a point. On the sides of this shadow there is a diverging shadow, the density of which decreases in proportion as it recedes from the former conical shadow: this is called the penumbra.

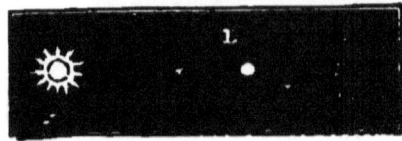

87. The quantity of the moon's disc which is eclipsed, and the same of the sun in a solar eclipse, is expressedly the number of digits: that is, the disc is supposed to be divided into twelve equally distant parallel lines. Then if half the disc be eclipsed the quantity of the eclipse is said to be six digits.

88. Seven is the greatest number of eclipses that can happen in one year, and two the least. If there be seven, five must be of the sun and two of the moon. If there be only two, both must be of the sun. There can never be more than three eclipses of the moon in the year, and in some places there are none.

CHAPTER XLI.

OF THE FIXED STARS, NEBULÆ, &C.

89. The bright points seen in the heavens at night are called fixed stars, because they have no apparent motion. They are supposed to be suns, each the centre of a system like our own, having planets revolving round them. The distance of the nearest fixed star, Sirius, is at least 19,000,000,000,000 miles. The fixed stars are known by their twinkling,—the planets by their steady light.

90. The stars, although called fixed, are not absolutely so, as astronomical observations have proved them to be in motion; but, from their immense distances, this motion is scarcely perceptible in half a century. It is supposed that there may not be really one fixed star in the heavens, but that on the principle of attraction, if one star be in motion, all the others must be affected thereby.

91. Stars are divided into six classes, according to their magnitudes, from 1 to 6. The largest are called stars of the first magnitude, and the smallest those of the sixth. Telescopic stars are those that cannot be seen with the naked eye.

92. To know the situation of the stars, they have been divided from time immemorial into various groups, to each of which some imaginary figure has been given. Each cluster so distinguished, is called a constellation; and these constellations, through which the sun appears to pass, and which the earth in reality passes over, during the year, are called the *twelve signs of the zodiac.* There are 93 constellations—35 in the northern hemisphere, 12 in the zodiac, and 46 in the southern hemisphere.

93. PERIODICAL STARS are such as appear and disappear. In 1572, Cornelius Gamma discovered a new star in Cassiopia, which dissappeared in 16 months. One appeared to Tycho Brahe, and in 1604 one appeared to Kepler. These stars are cupposed by some to turn slowly on their axis, and to have one dark side—but this is a very simple hypothesis. In such matters, it is better at once to confess our ignorance.

THE MILKY WAY, OR VIA LACTEA.

94. The milky way is that part of the heavens which appears of a light hazy colour at night. It is occasioned by myriads of telescopic stars. Dr. Herschel, on applying his great telescope to various parts of the milky way, found it to contain such increased numbers as to defy calculation. 116,000 passed through the field of view of his telescope in a quarter of an hour.

NEBULÆ.

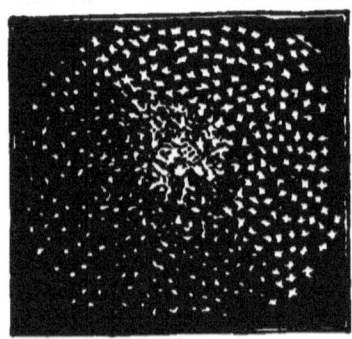

95. There are besides various shining spots in the heavens. These are called Nebulæ. The most considerable is one midway between the two stars, in the blade of Orion's sword, discovered in 1656 by Huygens, and that of Andromeda; others may be descried in Hercules, Saggitarius, and and Serpentarius.

96. Dr. Herschel made catalogues of 2,500 nebula. Some of these form a round compact system; others are more irregular, and some long and narrow. He considers them to form independent systems among themselves, and that our system is a part of a grander system, which includes the whole of the milky way; and that all the stars of the first, second, and third magnitude, belong to that vast cluster and shoal of stars.

97. Light which passes from the sun to us in eight minutes, would be no less than an hour and a quarter in passing from the nearest fixed star to our earth; and a cannon ball flying at the rate of 20 miles in a minute, would be 760,000 years performing the same journey. Sound, which moves at the rate of 13 miles in a minute, would be 1,128,000 years in reaching us from thence. According to the assumed distance of some of the fixed stars, their light must have left them at least 40,000 years ago.

98. In contemplating this wonderful starry host, which is but a mere speck to the infinity which lies beyond them, we are overpowered with wonder and admiration. But consider them and the millions whose light has not yet had time to reach our earth, through the immensity of their distance, but as a few particles of dust floating on the sunbeams, when put in comparison with that Eternal Spirit whose care at first created and still sustains them in their courses, calm, regular, and harmonious, but who, from this immeasurable height of majesty and glory, looks down upon the humble heart and shines on it,

"As shines a sunbeam in a drop of dew."

COMETS.

99. COMETS are moving masses of transparent fluids or vapours, and consist generally of a large, splendid, but ill-defined, cloudy mass of light, called the head or nucleus. From

this a long stream of light diverges, of various shapes, but generally broader, and more diffused at the greatest distance from the nucleus; this is called the tail of a comet.

100. Some comets are, however, quite destitute of this appendage, as was the case with the comet of 1682, described by Carsine to be as round and as bright as Jupiter; while that of 1744 showed six tails spread out like a fan. The comet of Halley in 1835 had three tails, spread out in different directions, while that of another comet resembled a Turkish scymitar. The tail of the great comet of 1680 was estimated to be 180,000,000 miles in length.

101. The comet of 1811 was 10,900 miles in diameter, and its luminous projection 132,000,000 miles. The comet which returned in October, 1835, called Halley's Comet, from that astronomer having calculated its return within a few days of its appearance, was in 1531 of a bright gold colour, 1607 dark and leaden, 1682 bright, in 1759 dark and obscure. The orbit of this comet was 3,420,000,000 of miles long. Its period is about 75 years.

102. Many comets have no nucleus, and the smallest stars have been seen through them; in those with a nucleus the light called nebulosity is not in contact with the nucleus. In the comet of 1811, the nebulosity was 20,000 miles from the centre of the nucleus, which was 2,700 miles in diameter.

103. The perehelion distance of the comet, 1680, was 150,000 miles from the sun, with a velocity of 880,000 miles an hour, while its aphelion was 2,898,000,000 of miles; the period 575 years. Others are estimated to have an aphelion distance from 15,000 to 66,000,000,000 of miles. The comet of 1843 went within a comparatively small number of miles of the centre of the sun, and had a tail of upwards of sixty degrees almost coincident with the plane of the horizon. It was at first taken for the zodaical light.

104. THE ZODIACAL LIGHT is a cone of light of a triangular form. It is brightest after sunset, about March and April, and before sunrise about September and October. It is supposed to arise from the sun's atmosphere.

CHEMISTRY.

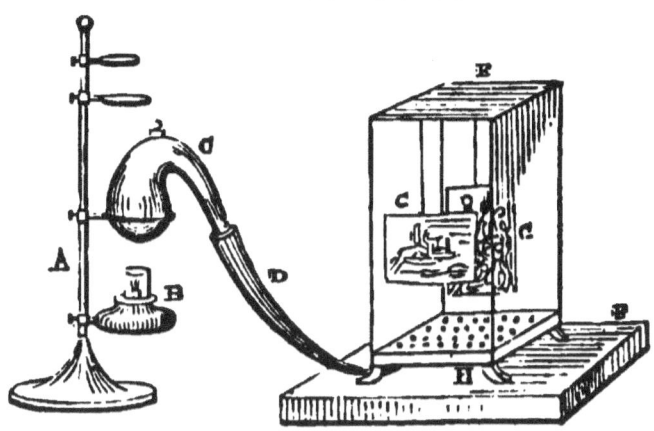

CHAPTER XLII.

DEFINITION—ITS CONNECTION WITH THE ARTS—SYNTHESIS AND ANALYSIS—METHODS OF INDUCTION—OBSERVATION, ANALOGY, AND EXPERIMENT—CHEMICAL AND MECHANICAL CHANGES—NATURE OF A CHEMICAL UNION—CHEMICAL COMBINATION—CHANGES OF PROPERTIES ATTENDING CHEMICAL COMBINATION—CHEMICAL DECOMPOSITION, ETC.

1. CHEMISTRY is the science which enables us to discover the *ingredients* of which bodies are composed, to examine the *compounds* formed by the *combination* of these ingredients, and to investigate the *nature* of the power which occasions these *combinations*.

2. CHEMISTRY has an intimate connection with all the arts of civilized life. *Dyeing, bleaching, tanning, glass-making,* and the *working* of *metals,* are chemical operations. In *agriculture, baking, brewing,* and *distillation,* a knowledge of it is of great importance; and in *medicine* it is indispensable. In short, Chemistry is everywhere: its laws preside over the most interesting phenomena of nature, and the most impor-

tant of those operations by which man supplies himself with the necessaries, comforts, and luxuries of civilized life.

3. CHEMISTRY has two ways of becoming acquainted with the internal structure of bodies, ANALYSIS and SYNTHESIS—in other words *decomposition* and *combination*. By the *former* process it *separates* the component parts of a compound body; by the *latter* it *combines* the separated elements so as to form anew the decomposed body and prove *the correctness of the former process*.

4. These two methods of induction depend upon a complete knowledge of the two powers by which all bodies in nature are set in motion, viz., ATTRACTION and REPULSION, which, as before stated, produce a great number of sensible phenomena, and a multitude of combinations which change the nature and property of bodies.

5. The foundations therefore of chemical philosophy are OBSERVATION, ANALOGY and EXPERIMENT. By *observation* facts are impressed on the mind; by *analogy* similar facts are connected; by *experiment* suppositions are verified and new facts discovered. Thus *observation* guided by *analogy* leads to *experiment*, while *analogy* confirmed by *experiment* becomes *scientific truth*.

6. DIFFERENCE BETWEEN CHEMICAL AND MECHANICAL CHANGES.—When a flint is broken by a hammer, wood split by a wedge, or a heavy body raised by a lever, the nature of the substance operated on still remains the same—its position or shape is only altered—the wood is still *wood*, the stone is still *stone;* but if the wood be exposed to a strong heat in an iron vessel almost closed, it *no longer remains the same*, but is converted into *charcoal, tar, vinegar*, and a *quantity of gasses*. The wood is no longer wood, having been resolved into its constituent principles—this is *chemical change*.

7. *Mechanical action* then is attended with sensible (apparent) motion, or change of *position*, or of *form*. *Chemical action* is not necessarily accompanied with sensible or apparent motion, but is attended by a change in the nature and properties of the substances on which we operate.

8. CHEMICAL COMBINATION.—If we pour a little oil into a phial, and add some water to it, these fluids will not combine together, however much they may be shaken, for when the

motion is over the oil will rise to the surface. If a small quantity of liquid ammonia be now added, and the mixture shaken as before, they will chemically combine ; for the oil and the ammonia, having a strong *affinity* for each other, *unite*, and a substance is formed which we call soap.

9. CHEMICAL DECOMPOSITION.—The oil and the ammonia in the above case combine because there is a strong affinity between them; but if another substance be introduced into the bottle having a stronger affinity for one of them than they have for each other, the compound will be decomposed. This may be effected by adding very carefully a small quantity of sulphuric acid (oil of vitriol) to the mixture. The acid has a stronger affinity for the ammonia than the ammonia has for the oil, and it will therefore leave the oil and combine with the acid, and the oil will again swim on the top.

10. DIFFERENCE BETWEEN A MECHANICAL AND CHEMICAL COMBINATION.—If a few iron filings are mixed with clay and a little water into a paste, they will combine together ; but this is simply a *mechanical combination*, for we shall not only be able to see the particles of iron by themselves, and the particles of clay by themselves, but we could pick out *mechanically* the particles of iron, and separate them from the clay. But if we take a *transparent saturated solution* of Epsom salts, and pour into it a like solution of caustic potash or soda, the two mixtures will unite and convert the mass into a form nearly solid—and by no mechanical means could we separate the Epsom salts from the potash, which as they have *chemically united*, can only be *chemically disunited*, or (technically) *decomposed* by some process analogous to the former example.

11. CHANGE OF PROPERTIES ATTENDING CHEMICAL COMBINATION.—When two substances combine chemically, there is generally a *change of properties* accompanying the combination. The compound produced has properties materially *different* from those of the substances which compose it. *Aquafortis*, a poison so corrosive that the least drop of it burns and corrodes the flesh, when combined with *potash*, a substance also of a caustic and corrosive nature, produces the *harmless substance* salt-petre ; and, *vice versa*, when *sulphur*, a *mild* and tasteless substance combines with *oxygen*, a gas also possessing

no acrid qualities, sulphuric acid (oil of vitriol) is produced, the most powerful of all acids, and one of the most acrid and corrosive substances known.

12. When we dissolve a piece of resin in spirit of wine we have an example of chemical union, for the resin will disappear and the spirit and the resin will unite in a light brown fluid; but if this mixture be poured into water, the resin will be reproduced, and fall to the bottom of the vessel as a white visible powder—the spirit of wine and the resin are *separated*—the mixture is *decomposed*. If a solution of sugar of lead (acetate of lead) be made with water, and a piece of metallic zinc be suspended in it, the zinc displaces the lead, which re-appears in the metallic state, separated in a beautiful crystalline form called the lead tree. Before the action we have *metallic zinc* and a *solution of acetate of* LEAD; after the decomposition we have *metallic lead* and a solution of acetate of ZINC.

13. By decomposition, *spirit of salt*, an acrid poisonous acid, is extracted from those wholesome matters, *salt* and *water*; and ammonia or hartshorn is separated from spirit of salt, with which it is combined in sal ammoniac.

14. It is by *decomposition* that wood or coals, when they are burnt, are separated into a light transparent invisible gas, which with smoke flies up the chimney, from the grey or white ashes which remain in the grate; and it is by *decomposition* that our own bodies, when the vital principle ceases, shall be resolved into their elements to be combined afresh into new compounds.

CHAPTER XLIII.

OF THE ELEMENTS OF BODIES.

15. Although there seems to be an almost infinite variety of substances in nature, yet when chemically examined they are found to be composed of but few distinct materials; these, which amount to 55* in number, are called elementary substances.

16. These elementary substances are arranged into two grand divisions, under the name of METALLIC and NON-METALLIC elements, and are subdivided as follows:—

* Some chemists do not reckon silicum an element, hence the number of elementary bodies is sometimes stated to be 54 only.

LIST OF ELEMENTS.

Non-Metallic Elements.

Oxygen.	Chlorine.
Hydrogen.	Iodine.
Nitrogen.	Bromine.
Sulphur.	Fluorine.
Phosphorus.	Boron.
Carbon.	Silicon.

Selenium.

Metallic Elements.

Potassium.	Copper.
Sodium.	Uranium.
Lithium.	Mercury.
Barium.	Silver.
Strontium.	Palladium.
Calcium.	Rhodium.
Magnesium.	Ruthenium.
Aluminium.	Iridium.
Glucinum.	Platinum.
Yttrium.	Gold.
Zirconium.	Osmium.
Thorium.	Titanium.
Cerium.	Columbium.
Lanthanium.	Pelopium.
Didymium.	Niobium.
Erbium.	Norium.
Terbium.	Tellurium.
Manganese.	Tungsten.
Iron.	Molybdenum.
Cobalt.	Vanadium.
Nickel.	Chromium.
Zinc.	Antimony.
Cadmium.	Arsenic.
Lead.	*Rubinium.*
Tin.	*Cæsium.*
Bismuth.	

CHAPTER XLIV.

OF OXYGEN, HYDROGEN, AND NITROGEN.

18. The word OXYGEN is derived from the Greek word οξυς sour, and γεννειν to generate; because when first discovered by Priestly in 1774, it was imagined that every *acid* contained it; but now it is known that many acid bodies contain no oxygen

19. OXYGEN is the basis of *vital air*, and one of the constituents of *water*. It is the chief support of *life* and *heat*, and performs an important part in most of the changes which take place in the mineral, vegetable, and animal kingdoms. It forms $\frac{8}{9}$ths of water, $\frac{1}{5}$th of air, about $\frac{1}{2}$ of all earthy matters, and exists largely in vegetables and animals.

20. Oxygen can only be obtained in the form of GAS, yet in combination with various substances, it may be either *solid* or *liquid*. The tarnish of silver and the rust of iron are only compounds of those metals with the oxygen of the atmosphere.

21. Oxygen is invisible, has neither taste nor smell, but its specific gravity is 1.1026; air at 60°, and barometer 30 inches, being 1.000. The weight of 100 cubic inches is 34.109 grains. Of 100 parts of water it absorbs 3.7 parts by measure; and when separated from a compound, by the electric energy, it always appears at the *positive* pole.

22. Oxygen may be obtained from the peroxide or black oxide of *manganese*, the red oxide of *lead*, the peroxide of mercury, nitrate of potash, and chlorate of potash. It is procured from the black oxide of manganese, by heating that substance in an iron retort, when from every 88 grains 18 grains are expelled in the gaseous form.

23. Oxygen is the active agent in the air. In respiration or breathing it is diminished *one-third* in quantity, and is either converted into *carbonic acid*, by combining with carbon in the lungs, or is absorbed and retained in the lungs, and carbonic acid is given off; or both may take place, as it is not determined which is the true theory of respiration.

24. *In germination* oxygen converts the carbon of the seed

into carbonic acid, and in the *respiration of plants* it plays the same part as in the *breathing of animals*. In combustion, or *burning*, oxygen is chemically united with the burning body. Its effects on metals is to convert them into oxides.

25. Any *compound* of oxygen, not of an acid nature, is called an *oxide*. When a metal is in the state of a rust, as seen on iron, or as seen in the tarnish of silver, or in the green verdigris of copper, it is said to be oxydized. To deoxydise is to deprive a body of oxygen.

HYDROGEN.

26. This term is derived from the Greek words υδωρ water, γεννειν, because it is one of the ingredients of water. It is the lightest of all known substances, weighing only about $\frac{1}{12}$ of the same bulk of *air*, $\frac{1}{11188}$ of the same bulk of *water*, and $\frac{1}{247738}$ of the same bulk of *platinum*, the heaviest body known. From its extreme lightness it has been sometimes used to fill baloons.

27. HYDROGEN GAS is found in a natural state in marshes, or stagnant waters, in hot weather, and may easily be obtained from them. If a bottle be filled with water, and a funnel put in, and both held downwards in a ditch, and the mud stirred at the bottom, the gas rises into the bottle and displaces the water. The ignis fatuus, or will o'wisp, originates from decayed vegetables, and the decomposition of pyritic coals. It consists generally of hydrogen combined with carbon, and perhaps occasionally with phosphorus or sulphur.

28. Hydrogen gas *does not* support combustion, as a lighted taper, if put into it, will go out, nor can it be *respired* for more than a few seconds; *sounds* in it become *acute*. It is exploded by a burning body, when *two* parts are mixed with *ten* or *twelve* of air.

29. Hydrogen gas is made by putting iron filings on small pieces of zinc into a bottle, with sulphuric acid diluted with four parts of water. An ounce of zinc with water and acid yields 676 cubic inches of hydrogen gas; an ounce of iron 782 inches.

30. When hydrogen is burned, *water* is the only product of its *combustion*, and the *decomposition* of water and its subse-

quent *re-formation* may be shown by the following easy experiment.—

"Add gradually one ounce of sulphuric acid to four ounces of water, in a large phial containing some iron filings. The temperature of the mixture will be so much raised by the union of the water with the acid as to enable the iron to decompose a part of the water. If a hole be neatly made through a cork, which fits the mouth of the phial, and a piece of tobacco-pipe be fitted into it, and the whole cemented into the phial with a piece of resin and bees-wax, the hydrogen gas as separated from the water will pour in a continued stream through the pipe, and may be set on fire by the flame of a candle brought in contact with it. The gas will continue to burn with a blue lambent flame as long as the decomposition goes on. This shows that the gas is *really* hydrogen, and that it arises from the *decomposition* of water."

"To *re-form* this gas into *water* hold a glass bell over the flame as it proceeds from the pipe; as the hydrogen burns it unites with the oxygen of the atmosphere, and the union of the two gases produces water, which will be seen to deposit itself like *dew* on the *inside* of the glass. This dew is the *water* which has been formed by the *combustion of hydrogen*."

31. Water can be decomposed in other ways. If forced through a tube over red hot charcoal it will be resolved into its elementary compounds. It may also be so resolved by that form of electricity called galvanism, and the two gases of which it was formed can be collected by means of a proper apparatus. Nature in many of her operations decomposes water. Fish and cold blooded animals are endowed with the same power. *See Illustration of Galvanism.*

32. OXIDE OF HYDROGEN, or WATER, consists of 8 parts by weight of oxygen, and 1 of hydrogen. It congeals at $32°$ F., and boils at 212. A cubic inch of distilled water, at the temperature of 60 weighs 252,458 grains. When formed into steam by boiling it expands into 1,696 times its former bulk. A pint measure of distilled water weighs a pound, consequently a cubic foot of water about 1,000 ounces or $62\frac{1}{2}$ pounds. The student should be careful to bear in mind the *specific gravity* of water, because this is always taken for unity in the measure of the specific gravity of every other substance.

33. MINERAL WATERS. In some situations water becomes so strongly impregnated with various matters that its taste and chemical qualities are greatly changed. Such waters are termed *mineral* or medicinal waters; of these there are four kinds:—1, ACIDULOUS or CARBONATED waters, containing much carbonic acid: 2, SULPHUREOUS waters, containing sul-

phureted hydrogen : 3, CHALYBEATE waters, which contain iron : 4, SALINE waters, which hold in solution soda, potash, lime, magnesia, &c. HARD water is water that contains a more than usual quantity of any of the compounds of lime, &c. The term *soft* is applied to water which either contains no earthy matters in solution, or so small a portion that it is not noticed.

NITROGEN.

34. NITROGEN is so called from its being *one* of the elements of *nitre*, and it is also called azote, or the *life-destroyer*. It is a clear inodorous gas, neither supporting combustion or respiration. It may be obtained by burning phosphorus under a close vessel, the oxygen being fixed by the combustion, and the residue, *nitrogen gas*, containing a very small proportion of carbonic acid and phosphorus in vapour.

35. Priestly discovered nitrogen to be the other constituent of the atmosphere, in which it acts as a *damper*, and prevents combustion and respiration from going on *too fast*. Did the air consist of oxygen *alone*, the least spark of fire would kindle the whole vegetable world into a flame, and in a few hours reduce them to ashes. Even metals would be set on fire, and man and other animals would not escape the general conflagration.*

36. NITRIC OXIDE.—This is a clear inodorous gas, and may be prepared by putting a few pieces of copper into a retort, and pouring on them nitric acid, when the gas will be evolved on the application of a gentle heat.

37. NITROUS OXIDE, or laughing gas, is composed of one atom of oxygen, and one of nitrogen, and is named *prot*-oxide from the Greek word πρωτυς, meaning first, because it is the first or smallest combination of oxygen with nitrogen. It

---

* Dr. Drummond in relation to this gas says:—" *Oil of Vitriol* is a strong corrosive poison, and a small quantity of it swallowed in its concentrated state would prove fatal; but when diluted, and consequently weakened by the addition of a large quantity of water, it can not only be swallowed, but is a pleasant drink and an efficacious strengthener of the stomach. Now the atmosphere forms a parallel case; the oxygen is too strong by itself, but every 20 parts being diluted with 80 parts of nitrogen, or azotic gas, the mixture forms the mild and grateful air on which our life depends."

cannot be made by mixing oxygen and nitrogen together; but it is procured by distilling nitrate of ammonia in a retort when the gas is given over.

38. When this gas is breathed it produces a curious effect on the human subject, exciting a disposition to laugh, jump, and dance, and gives rise to such a whimsical distortion of the features as to produce laughter from the beholders, if not from the experimenters, hence its name of *laughing gas*.

CHAPTER XLV.

CHLORINE—IODINE—BROMINE—AND FLUORINE.

39. CHLORINE.—The name of this element is derived from the Greek word (χλορος,) signifying green, because as a gas it is of a distinct greenish yellow colour. It will support combustion as phosphorus, and some of the metals in minute portions take fire when placed in it. But its most essential property is its power of *destroying vegetable colours*, hence its great importance in bleaching.

40. *To make chlorine.*—Take equal parts of per-oxide of manganese and common salt, and mix them in a wedgewood-mortar. Put about two ounces of this mixture into a retort, and also mix in a jug equal parts of oil of vitriol and water, stirring them well together with a glass rod; pour enough of the latter mixture into the former to make it into a thin paste, apply heat by means of a spirit-lamp, and the gas will be given off abundantly.

41. CHLORINE is used in great quantity for the purpose of whitening linen. Formerly the linen was exposed for some months to the process of washing, drying, scouring, and bleaching, by the agency of air and moisture; but by the employment of chlorine a piece of linen may be bleached in a few hours.

42. *Chlorine gas* destroys the volatile effluvia of putrefaction and infection. A table-spoonful of chloride of lime in a in a wine-glass of water, poured on a plate, purifies the air of

sick chambers, infected places, and removes smells from drains, &c. Its specific gravity is 2.5.

43. IODINE, or violet gas, resembles chlorine gas, and has some of its properties: it is a product from sea-weeds. Its specific gravity is 8.7, and 100 cubic inches weigh 262 grains. It changes vegetable *blue* to *yellow*, and $\frac{7}{1000}$ parts convert water into a deep yellow colour.

44. BROMINE is an elementary substance procured from sea-water, and in many particulars resembles *chlorine* and *iodine*. It is a liquid of a red colour, possessing a very disagreeable smell; it derives its name from the Greek word *bromos*, signifying *fetid*.

45. FLUORINE is a substance procured from fluor, or Derbyshire spar, and possesses the property of dissolving nearly every thing it touches. Etching on glass is effected by its dissolving the surface wherever it touches it.

CHAPTER XLVI.

OF CARBON AND ITS COMPOUNDS, OF SULPHUR AND OF PHOSPHORUS.

46. *Carbon* is in fact common CHARCOAL, divested of all its impurities; and its most striking property is that of crystallization, in which state it exists as the *diamond*, and this if *burnt* forms *carbonic acid*.

47. Carbon is found in large proportions, in bitumen, petroleum, and pit coal. It forms nearly the whole of all *vegetables*; it is also a component part of sugar and all kinds of wax, oils, gums and resins, and it enters into most *animal* and some *mineral* substances.

48. CHARCOAL is generally black, sonorous and brittle, very light, and destitute of taste and smell. It is a most powerful

The term *base* is sometimes applied to what is considered the leading ingredient in a compound substance; carbon is said to be the base of vegetables, lime the base of carbonate of lime, iron the metallic base of oxide of iron, sulphur the base of oil of vitriol, which consists of sulphur and oxygen.

antiseptic, has a great affinity for oxygen, is unalterable and indestructable by age, and if air and moisture be excluded, it is not effected by the most intense heat. Charcoal is the principal ingredient in gunpowder; it is also employed in purifying rancid oils, sweetening casks, &c.

49. When carbon is *combined* with iron, in *one proportion*, it forms CAST IRON; in *another*, STEEL; and in a third, PLUMBAGO, or GRAPHITE. In the first the proportion is about $\frac{1}{78}$; in the second $\frac{1}{780}$ part, and in the third it consists of 95 parts of carbon to 100 of iron.

50. CARBONIC ACID—This is composed of 6 parts of carbon with 16 of oxygen; but can only be exhibited in the form of *gas* called CARBONIC ACID GAS, each cubical inch of which weighs about half a grain. It may be collected in abundance from the surface of fermenting liquors, and may be obtained by pouring sulphuric acid upon a mixture of chalk or marble, and water. A cubic inch of marble contains as much carbonic acid in combination, as would fill a six gallon vessel when in the state of gas.

51. CARBONIC ACID GAS is *invisible* and *elastic*, and *twice* as heavy as common air. It will mix with common air, can be combined with water, and is destructive to animal life. It is found in nature in three different states—in gas, in mixture, and in *chemical combination*. As a *gas* it is about 1 part by measure of every 1000 parts of atmospheric air, and abounds in caverns and mines, where it is called the *choakdamp*. It is *mixed* with spa and other acidulous waters, and *chemically combined* with the *alkalies*, some *metallic oxides*, in *earths* and in *stones*, particularly in *chalk*, *lime-stone*, and marble.

52. Carbonic acid gas cannot be respired at all; whenever a draught of this gas comes in contact with the glottis, (the opening leading to the wind-pipe,) this aperture becomes quite closed up; but when about $\frac{1}{10}$ of this gas is present in the air it can be admitted into the lungs, and after being breathed for a minute causes a slight giddiness and tendency to sleep, which if suffered to go on would end in the sleep of death. Hence the danger of crowded rooms, the fumes of charcoal, gas, coke, &c. Its use as a chemical ingredient in the air is however of some importance, particularly in the decomposition of rocks, from which soils are formed.

53. CARBURETTED HYDROGEN is a compound of carbon and hydrogen, and is disengaged in various natural operations, as in the decay and decomposition of vegetables. Mixed with oxygen, or sufficient air, this gas forms an *explosive mixture*, which takes fire on the application of naked flame. In coal mines this property of the gas was once very destructive to human life; but Sir Humphrey Davy discovered, that if the flame of the miner was *surrounded by a cylinder of wire gauze*, that no explosion took place, hence his invention of the safety lamp, which enables the miner to have light without danger *(See Illustrations.)*

54. CARBON, therefore, whether we regard it in its most simple state, the diamond, or in that of *common charcoal*, is not only indestructible by age, but in all its combinations. In the state of *carbonic acid* it exists in union with earth and stones in abundant quantities, and although buried for thousands of years beneath immense rocks, is still carbonic acid; and no sooner is it disengaged from its dormitory than it rises with all the life and vigour of recent formation, forming as it were a type of human renovation. In the same manner the leaves that fall and rot upon the ground, apparently destroyed, are not lost; the oxygen of the atmosphere combines with their carbon, and converts them into carbonic acid gas, and this same carbon is absorbed by a new race of vegetables;

"Link after link the vital chain extends,
And the long line of being never ends."

SULPHUR.

55. SULPHUR is a simple combustible substance found native in a loose powder, either detatched, or in veins. It is met with in the neighbourhood of volcanoes, where it is deposited as a crust on stones contiguous to them. It can be prepared by exposing iron pyrites (*sulphuret of iron*) to heat; when part of the sulphur is driven off in vapour, and may be collected in water. When vaporized it condenses in small crystalline particles, called flowers of sulphur. It is *inflammable*, burning slowly with a pale blue flame.

56. Sulphur is found in connexion with *silver, copper, lead, antimony,* and *iron.* If is a *negative* electric; specific gravity 1.99; it melts at 240° and may be cast to 280°; it thickens

by evaporation at 320°, and at 428° becomes a soft paste; at 550° it boils, evaporates and produces flower of sulphur.

PHOSPHORUS.

57. The real origin of phosphorus is very obscure. It is a peculiar substance entirely of animal origin. It is now always procured by the decomposition of the phosphoric acid found in animal bones. It is *solid* and very combustible, burns at a very low temperature, 100° Fah., and absorbs in its combustion half as much again of oxygen as its own weight. When burnt in oxygen gas it gives forth a flame of the most vivid and intense brightness. Surrounded by cotton, rubbed in powdered resin, it takes fire *spontaneously* under the receiver of an air pump when exhausted; and displays a very beautiful phenomenon on the gradual admission of the air. It is used in making match bottles for instantaneous light, by mixing *one* part of flowers of sulphur with *eight* of phosphorus. The light of the glow-worm is phosphorescent, and the extraordinary light seen upon the waves of the sea in warm latitudes is produced from millions of animalculæ possessed of a phosphoric nature.

CHAPTER XLVII.
THE METALLIC ELEMENTS.
NON-ALKALINE EARTHS.

58. The bases of the *non-alkaline* earths are Silic*um*, Alumin*um*, Thori*um*, Glucin*ium*, Zirconi*um*, Ittri*um*, and these earths bear the name of Silica, Alumina, Thorina, Glucina, Zirconia, and Ittria.

59. Before the discoveries of Sir Humphrey Davy not only were the alkalies, such as potash and soda, imagined to be simple bodies, but also the substances known by the names of earths. But this great philosopher proved that *potash*, *soda*, and *lithia* were *metallic oxides*, and subsequently it was ascertained that *baryta* was found to be the oxide of a metal since called *barium*, and *strontia* the oxide of a metal called by its

discoverer *strontium, lime* the oxide of *calcium*, and *magnesia* of *magnesium*. The five *non*-alkaline earths have been also found to possess a similar constitution, and are oxides of the metals to which their names refer.

60. SILICA and ALUMINUM are abundant. Silica exists in *sand, flint,* and is nearly pure in *quartz;* and the *agate, jasper,* and *opal* consist almost entirely of it. It is supposed to consist of 22 parts of *silicum*, and 24 of *oxygen*.

61. ALUMINUM is sometimes called argillaceous earth, as it gives to clays their peculiar character—softness, plasticity, and adhesiveness. It is very abundant, forming a considerable portion of most rocks and earths; and many precious stones, as the *sapphire, ruby* and *turquoise,* contain it. It consists of about 26 parts of the metallic base *aluminum*, and 24 parts of oxygen.

62. GLUCINA, ITTRIA, ZIRCONIA, and THORINA are very rarely met with, but they resemble the earthy matters in their proportions and composition. The EMERALD and BERYL contain glucina united with silica, alumina, and lime. Zirconia forms about 70 parts of 100 in the hyacinth; ittria constitutes above half of a rare mineral called GADOLINITE, and thorina forms about 58 parts in a 100 of a mineral called THORITE.

ALKALINE EARTHS.

63. The bases of the alkaline earths are *calcium, barium, strontium* and *magnesium*, and the earths are called *lime, baryta, strontia,* and *magnesia*.

64. These metals are prepared with great difficulty, have been seldom seen, and little is known of their properties. Lime is found in the *animal, mineral,* and *vegetable* kingdoms; it is usually obtained by exposing chalk or marble, which are carbonates of lime, to a strong heat; carbonic acid is expelled and lime in a pure state remains.

65. Pure *lime* mixed with sand and water forms *mortar*, which imbibing carbonic acid from the atmosphere as it dries, becomes lime-stone again, and thereby cements walls of brick and stone. Slacked lime is termed chemically, hydrate of lime.

66. If lime powder, from which the carbonic acid has been

expelled, be exposed to the intense heat of the union of oxygen and hydrogen at the poles of a voltaic battery, or at the current of a blow-pipe, it is converted into *calcium*, and during the combustion gives out so much light that a small piece of it, not larger than a pea, will illuminate a lighthouse. It is used in the same manner in the hydro-oxygen microscope. *See Illustration (Optics).*

67. Baryta is of a greyish white colour, and is remarkable for its weight, and strong alkaline qualities, such as destroying animal substances, turning blue vegetable colours green, and showing a powerful attraction for acids.

68. Magnesia is a soft white light earth, unalterable in the fire, and almost insoluble in water; combined with sulphuric acid it forms sulphate of magnesia, and is the most effectual antidote in case of poison by some mineral acids.

BASES OF THE ALKALIES.

69. The bases of the alkalies are *potassium, sodium,* and *lithium,* and their oxydes or earths are *potassa, soda* and *lithia.*

70. POTASSIUM is formed by exposing a hydrate of potash to a voltaic circle of 500 double plates of four inches, when the substance appears at the negative pole, oxygen being developed by the positive pole. It possesses very remarkable properties, and has an affinity for oxygen more powerful than any other metal. When it comes in contact with water it *takes fire,* and burns with a reddish or violet flame, moving rapidly over the surface of the water, amid explosions and the evolution of a quantity of gas.

71. POTASH is the matter which when in union with carbonic acid, forms the common potash or pearlash of the shops: these matters are carbonate of potash. The terms *potash* and *potassa* are synonymous. *Potash* consists of *potassium* in union with oxygen, and is characterized by being exceedingly caustic, having a great affinity for water, and also for carbonic acid, in virtue of which it is enabled to abstract water and carbonic acid from the air.

72. POTASSA is procured by abstracting the carbonic acid from the carbonate of potassa or lime. The lime and the carbonic acid form an insoluble mass; the potassa remains in

solution in the water in which the carbonic acid was dissolved When this water is expelled by boiling, the potassa appears as an oily liquid which becomes solid on cooling.

73. SODIUM.—Like potassium, sodium decomposes warm water with rapidity, but does not usually give rise to combustion, except the water be so thickened with gum that the globule of sodium cannot roll about. The most interesting compound of sodium is common salt, which consists of 36 parts of chlorine, and 24 of sodium. It is termed chemically *chloride of sodium*.

CHAPTER XLVIII.

REFERRING TO DIVISION II. OF THE METALLIC ELEMENTS.

(See Table of Elements.)

METALS.

74. The *metals* appear to be simple substances, and are distinguished from all bodies by a peculiar *brilliancy*, which is called *metallic lustre*, and by their *weight* or specific gravity, *malleability* and *ductility*.

75. Metals are found in the oldest rocks in veins and in fissures, seldom *pure*, but in a state of combination with other metals, and with *sulphur*, *oxygen*, and *acids*. Every metal possesses some specific difference from the others, and they are found under different forms, or geometrical figures of varied character. These peculiarities of the figures of the particles that compose them, are not destroyed when reduced by art to a metallic state. Thus when the surface of a melted metal begins to congeal, if the crust be broken, and the parts still in a fluid state be drawn off, the parts which have cooled will exhibit a regular metallic crystallization.

76. Metalliferous veins and fissures extend irregularly to unknown depths, and occur mostly in primary or secondary formations. The veins vary in width from an inch to 20 feet. Lime-stone is the most metalliferous of the secondary rocks, and lead and copper are the metals most usually found in it.

In density the order of the metals is platina, gold, silver, mercury, lead, copper, tin, iron, and zinc.

77. PLATINA, or platinum, was first ascertained, to be a perfect metal by Scheffer, a Swedish chemist, in 1752, and soon became subject to the experiments of all the chemists in Europe, who determined it to be equal in most of its properties to gold, but in others very superior. Platina is generally mixed with iron; it is malleable and ductile; it is commonly found combined with some one of four other metals, *Osmium, Irridum, Rhodium*, and *Palladium.*

78. GOLD is always found in a metallic state. It is generally met with in grains where it is seen in leaves and crystalline ramifications, adhering to quartz and other stone. It is not tarnished by air or moisture, and is dissolved by a mixture of nitre and sal ammoniac, called *aqua regia*. All the known parts of the earth afford gold, but the greatest quantity is obtained from the mines of Australia.

79. SILVER is found native, and also combined with *lead, copper, mercury, cobalt, sulphur*, &c. When found in a metallic state it appears in grains or leaves. There are several silver mines in various parts of the world, but the largest quantities come from Peru and Mexico.

80. With nitric acid silver forms a colourless solution, called nitrate of silver, which stains animal and vegetable substances with an indelible black colour; hence its use as a permanent ink, and to dye hair, whiskers, &c. The affinity of silver for other metals, and its solution in acid, are the properties on which plating and silvering depend.

81. MERCURY, called also quicksilver, always appears in a liquid state, and is sometimes found pure; but generally united to *sulphur*, when it is called *cinnabar*, or sulphuret of mercury; chloride of mercury, commonly called *calomel*, consists of 36 of chlorine and 200 of calomel; *corrosive sublimate*, called bi-chloride of mercury, consists of 200 of mercury and 72 of chlorine. The fine pigment, called *vermillion*, is prepared by fusing 200 of mercury with 32 of sulphur. A cubic inch of mercury at 62°, Bar. 30°, weighs 3425.35 grains.

82. The quicksilver mines of Carniola are the most productive in Europe, and yield 1,200 tons' weight yearly, with half the amount of cinnaber. The mercury is found in clay stones, and often issues 'rom the rock spontaneously.

83. LEAD.—This metal is very seldom found in a metallic state. It is chiefly combined with sulphur, when it is called *galena*, or sulphuret of lead. When exposed to heat, with access of air, it fuses and is oxydated at its surface. If this oxide be removed more is formed, and thus the whole may be converted into *grey oxide of lead*. This oxide when exposed to a strong heat is converted into a yellow oxide, called *massicot*. If this yellow oxide be exposed to a still more violent heat it assumes a beautiful red colour, and becomes *red lead*, or minium. *White lead* is a carbonate of the oxide of lead, prepared by exposing metallic lead in thin sheets to the vapour of acetic acid. *Sugar of lead* is acetate of the oxide of lead, prepared by steeping carbonate of lead, or litharge, in acetic acid.

84. IRON is the most universally diffused metal throughout nature, and is found in combination in every part of the mineral, vegetable, and animal kingdoms. It has *hardness, elasticity*, and *ductility*. It is very difficult to fuse, and soon rusts or oxydates when exposed to the action of the air or atmosphere. It burns brilliantly in oxygen gas.

85. IRON is fuzed from the ore in large furnaces, and made to flow into moulds of sand. It is then called cast iron, of which are formed stoves, pipes, cannon, and other articles. In this state it may be either of the three kinds, *white, grey,* or *black,* according as it contains a *larger* proportion of *carbon,* an *exact* proportion of *carbon* and *oxygen,* or a *larger* proportion of *oxygen.*

86. *Iron* is capable of being converted into a substance called STEEL, by exposing it to heat in contact with a carbonaceous substance, such as charcoal, for several days. The process is kept from the atmospheric air, and the iron when hammered at a red heat becomes steel.

87. The RED OXIDE, or PEROXYDE of IRON, is obtained by putting red hot iron filings into an open vessel, and agitating them till they produce the common red paint. It is this oxide which produces the red colour of bricks and clays. *Protoxide* of iron, united with sulphuric acid, forms SULPHATE OF IRON. The peroxide, united with gallic acid, forms the colouring matter of *ink*. The same oxide, united with ferrocyanic acid, forms *prussian blue*.

88. MAGNETIC PYRITES is a sulphuret of iron, 63 iron, and 37 sulphur. *Super-sulphuret* of iron (iron pyrites) is 47 iron and 53 sulphur. *Carburet of iron*, or iron and carbon, (plumbago), contains 9 parts of iron and 1 of carbon. The article, called *green copperas*, is a *sulphate of iron*.

89. COPPER is found native but in very small quantities. it is generally met with in the state of an oxide. Pure copper is of a red colour, *tenacious, ductile*, and very *malleable*. It soon rusts on an exposure to the air, a carbonate of the peroxide of copper being formed. NITRATE OF COPPER is copper dissolved with nitric acid. The SULPHATE OF COPPER, commonly called *blue vitriol*, is sulphuric acid and copper. *Verdigris* is acetic acid imperfectly oxidated with copper. It is prepared on a large scale by the action of copper on the refuse of the wine making process. CARBONATE OF COPPER is prepared by the action of *carbonate of soda or potash* on *sulphate* of copper.

90. Copper may be alloyed with most of the metals. As an alloy of silver it renders it more fusible, and is employed as a solder for silver plates. Copper when alloyed with tin forms bronze, a metal used for making bells, cannon, statues, &c. When alloyed by cementation with the oxide of zinc, called *calamine*, it forms *brass*.

91. TIN is generally found in combination with sulphur, lead and other substances. It is procured from the oxide of tin by the action of charcoal at a high temperature. It does not rust, or oxydate. It is the lightest of the metals. It is soluble in sulphuric acid. With muriatic acid it forms muriate of tin, of great use in dying. It alloys with other metals forming *solder*. With *lead* and *antimony* it constitutes *pewter*. Its compounds with oxygen and chlorine are used in calico printing for *fixing* certain colours, and for *producing* or *discharging* others.

92. ANTIMONY has a brilliant white colour, with a shade of blue; when melted it emits a white flame, while cooling it slowly crystallizes in octahedral crystals. It is very little changed by exposure to the air. Steam passing over it in a red hot state is so rapidly decomposed as to occasion a violent *detonation*. It is of great service in medicine, and is used as an alloy in printers' type, small shot, mirrors of telescopes, &c.

93. ARSENIC, or Arsenicum, is of a grey colour, brittle, and harder than copper. It is a highly inflammable metal, and proves extremely corrosive to animals, so as to become a violent poison. It soon loses its metallic lustre in the air, and becomes black. It is often found native; when combined with sulphur it is called *orpiment*. If thrown upon a hot surface it flies off in vapour, which possesses a garlic odour. The white arsenic of the shops is properly arsenious acid, and is a deadly poison.

94. ZINC is a bluish white metal, very *brittle* at certain temperatures, but at others both malleable and ductile. It is combined in various proportions artificially, forming valuable alloys. Brass consists of one part of zinc to four of copper; when more zinc than this is used, then compounds are generated, called Tombac, Dutch gold, and Pinchbeck.

95. COBALT is a metal of a reddish grey colour, *brittle, pulverizable*, and *magnetic*. It is as difficult to fuse as iron. It yields fine colours to the painter. A precipitate of cobalt, by a potash lye from nitro-muriatic acid and nitrate of zinc, yields a beautiful *green*. Powdered cobalt inflames if thrown into oxygenated muriatic acid or chlorine.

96. NICKEL is of a white colour, very hard and infusible, and always a constituent of meteoric iron. The oxide of nickel gives a fine hyacinth colour to glass, hence it is used in making artificial gems.

97. MANGANESE is a light grey metal, very *brittle* and *difficult to melt*. It absorbs oxygen from the atmosphere so rapidly as to change from a white to the red and black oxides, and to fall into powder. If thrown into water, it decomposes the water, becomes green, and it is found to have absorbed 0.15 of oxygen; if this be exposed to the air it turns brown, and is found to contain $\frac{1}{4}$th oxygen; in a native state the oxide contains four tenths, hence its importance in the production of oxygen gas.

98. PALLADIUM, RHODIUM, IRIDIUM and OSMIUM are found mixed with native grains of platinum. *Palladium* is named from the planet Pallas. *Rhodium* from Rhodon, a rose, because of the rose colour possessed by its compounds. *Iridium* from Iris, the rainbow, because of the variety of tints seen in its salts; and *Osmium* derives its name from *Osme*, smell, because its compounds have all a very strong odour.

99. Selenium is a new metal, obtained by Berzilius from the pyrites of Fahlun; its specific gravity is 4.32. Cadmium is another metal in union with zinc, with a specific gravity of 8.604, and is of a grey colour.

CHAPTER XLIX.

ACIDS AND SALTS.

100. Acids are either solid, liquid, or gaseous. They excite a peculiar sensation on the tongue, called sour. They change most of the blue, green, and purple juices of vegetables red, and combine with the earths or metallic oxides so as to form those compounds, called SALTS. Most of the acids owe their origin to the combination of certain substances with oxygen, which has been called the acidifying principle.

1. *Mineral Acids.*
Sulphurous, Sulphuric, Nitrous, Nitric, Muriatic, Carbonic, Fluoric, Boracic.

2. *Metallic Acids.*
Arsenious, Arsenic, Tungstic, Molybdic, Chromic, Columbic.

3. *Vegetable Acids.*
Acetic, Malic, Oxalic, Citric, Tartaric, Benzoic, Camphoric, Gallic, Succinic, Surberic.

4. *Animal Acids.*
Phosphorous, Phosphoric, Sebacic, Luccic, Lactic, Uric, Prussic.*

101. Sulphuric acid is obtained by burning sulphur and nitre in a close receiver over water; it then becomes, after being concentrated by boiling, what is vulgarly called oil of vitriol.

102. Nitric acid is what is sold in the shops under the name of aquafortis; it is obtained from nitrate of potash and

* We have preferred classing the acids thus, this classification being the most simple and easy of comprehension.

sulphuric acid by means of heat, and will dissolve all metals except gold and platina.

103. MURIATIC ACID, *vulgo* spirits of salt, is a peculiar acid, obtained from sea salt by means of sulphuric acid. According to Sir H. Davy, muriatic acid is a compound of hydrogen and oxymuriatic acid. Oxygenated muriatic acid is muriatic acid with an excess of oxygen.

104. CARBONIC ACID is produced from the combustion of carbon: it exists in the gaseous state, and in combination with the different bases it constitutes carbonates.

105. FLUORIC ACID is obtained from Derbyshire spar, and is separated from it by sulphuric acid. It has great affinity for silex, and engraving may be performed with it on glass, as etching is on copper by nitric acid.

106. BORACIC ACID is a concrete acid extracted from soda and borax, with which it is combined, by sulphuric acid, and appears in the form of shining scales: it may be decomposed by burning in contact with it *in vacuo* potassium; the potassium attracts the oxygen from the acid, and leaves its basis in a separate state; by burning this in oxygen gas, its light is extremely brilliant, and the result is boracic acid.

107. MOLYBDIC ACID is a pale yellow powder, procured from the sulphuret of molybdenum, by distilling nitric acid off it repeatedly, till the sulphur and metal are both acidified.

108. CHROMIC ACID is compounded of chromium and oxygen, and forms an orange-coloured powder: it has a hot metallic taste, dissolves easily in water, and when mixed with different saline solutions, it assumes a variety of beautiful colours.

109. ARSENIC ACID is composed of arsenic and oxygen. TUNGSTIC ACID is a tasteless yellow powder, composed of oxygen and tungsten. ACETIC ACID is principally obtained from saccharine liquors which have undergone the acetous fermentation; it is known in commerce by the name of vinegar. OXALIC ACID is a peculiar acid found in sorrel, in combination with potash, and its base is in all saccharine substances; it is much used by the calico printers, and is a deadly poison. TARTARIC ACID is found in the cream of tartar of commerce. CITRIC ACID is found in the juice of lemons, &c. MALIC ACID is found in the juice of apples

SALTS.

LACTIC ACID is prepared from milk. GALLIC ACID is found in the galls of commerce. BENZOIC ACID is prepared from gum Benzoin. Succinic acid is obtained from amber; CAMPHORIC ACID from camphor; Suberic acid from cork; LACCIC ACID from white lac. PRUSSIC ACID is a peculiar acid composed of hydrogen, nitrogen, and carbon, and so strong a poison as occasionally to destroy life upon merely coming in contact with the skin; it is supposed to act by de-oxidizing the blood. SEBACIC acid is produced from animal fat or tallow. The URIC ACID is found in human urine, and is composed of carbon, nitrogen, hydrogen, and oxygen.

110. These acids are such powerful agents in a variety of chemical changes which take place in nature and in the arts, that it is of the utmost importance to acquire a knowledge of the mode in which they operate. There are two ways in which the acids produce changes in the substances with which they are brought into contact: in some cases they effect a union with these substances, and become a part of the new compound, without having themselves undergone any decomposition; in others, they become partly decomposed by affording a part of their oxygen to the bodies on which they operate: the formation of saltpetre, or nitre, with the addition of nitric acid to potash, is an instance of the first of these cases, and the action of the same acid upon iron will exemplify the latter.

OF SALTS.

111. The word salt was originally confined to chloride of sodium, our common salt; the term is now applied to all the compounds which the acids form with alkalies, earths, and metallic oxides. The number of species of salts is not ascertained; the number of distinct salts is constantly increasing. Chemists have agreed to denominate the salts from the acids they contain: thus all substances which are compounds of metallic oxides, earths, or alkalies, with sulphuric acid, are called *sulphates*; with muriatic acid, *muriates*; with nitric acid, *nitrates*; with carbonic acid, *carbonates*, &c., but sometimes an adjective termination is given to the word, which expresses the base of the acid, as calcareous salt, instead of salt of lime, ammonical salts instead of salts of ammonia

112. In conformity with this plan, the saline compound formerly called Glauber's salt, is now called *sulphate of soda*, for it is a combination of sulphuric acid and soda; what was called gypsum, or plaster of Paris, a compound of lime and sulphuric acid, is now called *sulphate of lime*, and what was called green copperas, is now called *sulphate of iron*, that substance being a compound not of copper, but of iron and sulphuric acid

113. Some salts are formed with acids not fully oxygenated, as the sulphurous and phosphorous acids; all salts therefore that are composed with acids, ending in *ous*, take an ending in *ite* instead of *ate*, as sulphite of lime, or phosphite of potash. When a salt is found to contain an excess of acid, the preposition *hypo* is generally prefixed to its name, but when it does not contain a sufficiency of acid to saturate the base, the preposition *sub* is added; thus we say, hypotartrate of potass, and sub-borate of soda.

TOXICOLOGY.

114. This is the science of poisons and their antidotes, and its name is derived from *toxon*, a bow, evidently relating to poisoned arrows. Oil of vitriol is a chemical poison, and kills by destroying the coats of the stomach. To this magnesia is the best antidote, but it must be mixed with a *small* quantity of milk. Oxalic acid is a poison, but this combined with lime is inert; therefore any of the carbonates of lime, such as chalk, whiting, or even mortar of the cieling, may be given. Epsom salts is the best antidote for poisoning by salt of baryta. Poisons from any of the preparations of copper or of mercury, such as corrosive sublimate, may be neutralized by the white of eggs beat up in water. There is unfortunately no remedy for poisoning by arsenic: one test is lime water, which will separate it—a *white precipitate* remains; if a drop or two of nitrate of silver, and a single drop of hartshorn be added to a solution of it, a *yellow* precipitate falls: if it be tested in like manner with a small quantity of sulphate of copper and hartshorn, a *green precipitate* falls

CHAPTER L.

OF HEAT.

115. OF HEAT.—Heat, in a latent state, exists in all substances with which we are acquainted, though it combines with different substances in very different proportions. Sensible heat is an effect produced upon our organs by contact with surrounding bodies. When we touch a warm body, a portion of its heat leaves it to pass to us; when we touch a cold body, a portion of our heat passes to it.

116. Heat may be considered under two heads—free or radiant heat, and specific heat. Free heat is that which is not in any way combined with any other body; and specific heat is that which is fixed in bodies.

117. Free heat has the power of expanding bodies, and causes them to occupy a greater space,* and if the temperature be sufficiently great, they become burnt, fused, or melted. Different bodies, subjected to the same temperature, have different capacities for heat, and this is called their specific heat.

---

* If an iron ring be made of such a size that an iron ball in a cold state can just slip through it, and the ball be then made red hot, its bulk will be so much increased that the dimensions of the ring will be too small for

Exp. Take a pound of lead, a pound of chalk, and a pound of milk, and having placed them in a hot oven, they will be gradually heated to the temperature of the oven, but the lead will acquire this temperature first, as having the least specific heat, the chalk next, and the milk last. To ascertain that a less quantity of heat enters into lead than into the other bodies to raise it to the same temperature, it will only be necessary to place these three bodies in equal quantities of water of the same degree of coldness, when it will be found that the water in which the lead has been immersed will be the least heated, that of the chalk next, and the milk most of all. We must not consider that this proceeds from the chalk and milk having greater bulk than the lead, as equal weight must always contain equal quantities of matter, though their dimensions may be different.

118. Latent heat is that portion of heat which is insensible in all bodies. When a body changes from solid to liquid, or from liquid to vapour, it immediately absorbs a quantity of heat which becomes sensible, when the body from which it has been transferred reverts to its former state.

Exp. Mix with snow muriate of soda, and it will be found, by immersing the thermometer into the mixture, to be many degrees below the freezing point; expose this to the heat of the sun, and the thermometer will be observed to descend till all the ice is melted; when, however, all the ice is melted, the thermometer will begin to rise: the reason is, that the mixture absorbs the heat without its temperature being raised.

119. As solids converted into fluids absorb latent heat, so fluids converted into solids give forth their heat.

---

its passage. The diameter of the ball is, therefore, greater than before, being expanded by the heat which entered it.

The engraving at the end of this chapter represents an instrument for measuring the comparative expansion of solid bodies: it is called a pyrometer. Expansion is denoted by the pressure of the heated body against a part of the machinery, which acting upon the other parts, the whole receives a multiplied motion, which is marked by the index. A is a metallic rod undergoing expansion; B, a treble lamp filled with alcohol; C, the wheel acted on by the least pressure; D, the index; E, the stand.

The art of making watch-glasses depends upon the employment of heat as a power which expands glass unequally. A glass globe is blown of sufficient size to permit five glasses to be cut from it. When the globe is cold, a red hot wire is run round, marking the size of the watch-glass, which drops out of its place.

Exp. Take a solution of sulphate of soda, made very strong, and put it in a bottle hot, which cork immediately; observe that this will not then crystallise; when it is cold uncork the bottle, and it will be seen that on the admission of air the salt will suddenly crystallize, and give out its latent heat, so that the bottle will be warm.

120. All bodies do not radiate alike, for their surfaces have great power of promoting or retarding the radiation.

Exp. If a cubical canister, four inches square, of polished tin, be filled with boiling water, and placed at three feet distance from a concave tin reflector, which shall have the non-graduated ray of a differential thermometer in the focus, the quantity of radiated heat will be denoted by the rising of the fluid in the graduated ray to about 120°. But if the canister be brushed over with a mixture of size and lampblack, the thermometer will rise to 300°. The experiment may be varied by coating the canister with different substances, which, according to their surfaces, textures, and consistencies, radiate more or less heat. Hence may be understood why a bright silver or tin tea-pot makes better tea than a black stone or wedgwood one. (See engraving at the head of this chapter.)

121. COMBUSTION.—Combustion is a process by which certain substances combine with oxygen gas, absorb its base, and suffer its caloric to escape in the state of sensible heat, accompanied with light and flame.

122. The *simple combustibles* are, hydrogen, sulphur, phosphorus, carbon, and all the metals except gold, silver, platina, and mercury. Compound combustibles are formed by the union of two or more of the simple combustibles. Hydrogen and carbon, intimately united in the capillary tubes of vegetables, form *bitumens, oils, resins, coal, &c.,* which are *compound combustibles.*

123. The supporters of combustion are oxygen, atmospheric air, nitrous oxide, and some others. When combustion takes place, the combustible combines with the supporter of combustion, and forms a compound which is usually carbonic acid and water.

124. When a body is fully burnt it is said to be oxydized, and cannot be rendered combustible again but by depriving it of its oxygen, with which it combined in its former combustion. The chemical term given to burnt bodies is that they are oxydized; that is, changed into oxides. Thus the process of combustion merely decomposes a body, and sets its component parts at liberty to separate from each other, and from other new and varied combinations, and nothing less than consummate wisdom

could have devised so beautiful a system; thus we are led to refer every thing to design, unerring intelligence, combined with infinite goodness.

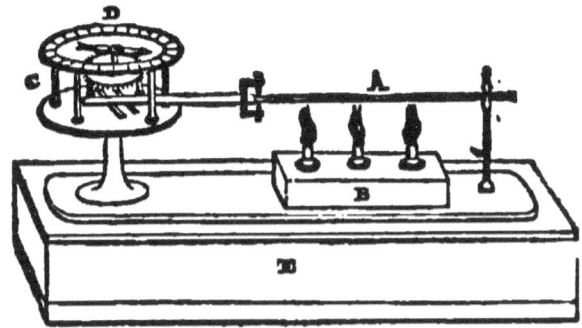

Instrument for measuring the comparative expansion of solid bodies.

ELECTRICITY.

CHAPTER LI.

DERIVATION—ELECTRIC ACTION—ELECTRICAL MACHINE—
THEORIES OF ELECTRICITY—EXCITATION—DISTRIBUTION
—ATTRACTION AND REPULSION—INDUCTION—LEYDEN JAR
—ELECTROMES—UNIVERSAL DISCHARGER—DE LUC'S CO
LUMN—ANIMAL ELECTRICITY.

1. The term electricity is derived from the Greek word *electron*, amber, because electrical attraction was first discovered from this substance. Electricity, therefore, primarily treats of the phenomena and effects produced by the friction or rubbing together of certain bodies called *electrics*. These consist of *glass, amber, resinous matters, silk, hair, wool, feathers, various vegetable substances*, and *atmospheric air*.

2. To show the nature of electrical action, rub a piece of amber or sealing-wax on the coat, and it will be found, while warm by the friction, to attract light bodies, such as straws or small pieces of paper. If a clean glass tube be rubbed several times through a woollen or leather cloth, and presented to any small substances, it will immediately attract or repel them; and, if a poker, suspended by a dry silk string, be pre

sented to its upper end, then the lower end of the poker will exhibit the same phenomena as the tube itself, which shews that the electric fluid passes through the metal; but if for a metallic body a stick of glass or sealing-wax be substituted, these phenomena will not occur, which proves that the electric fluid does not pass through these substances.

3. By this it will be perceived, that besides the class of bodies called *electrics*, there is another which are called *conductors*. These bodies cannot be excited themselves, but have the power of transmitting this electric influence through them. These bodies comprise *all the metals, semi-metals*, and *metallic ores*. The *fluids* of *animal bodies*, water, and other fluids, *except oil, ice, snow, earthy substances, smoke, steam*, and even a vacuum.

4. When any electrified conductor is wholly surrounded by *non-conductors*, so that the electric fluid cannot pass from the conductor along conductors to the earth, it is said to be insulated: thus—the human body is a *conductor* of electricity; but if a person standing on a glass stool be charged with electricity, the electric fluid cannot pass from him to the earth, and he is said to be *positively electrified*, because he has more than his natural share; he is also *insulated*; and if he be touched by another person standing on the ground, sparks will be exhibited at the point of contact, where the person touching will feel a sharp pricking sensation.

5. In order to illustrate certain remarkable facts in this science, attention must be directed to the figure:—B is supposed to be a small piece of cork, or the pith of wood, which is suspended from a stand by a dry silk thread. Having rubbed an electric, for instance, a dry rod of glass, and presented it to B, the ball will be instantaneously attracted to the glass, and will adhere to it. After they remain in contact for a few seconds, if the glass be withdrawn, without being touched by the fingers, and again presented to the

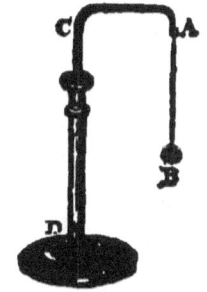

ball, the latter will be *repelled* instead of attracted, as in the first instance. By being touched with the finger, the ball can be deprived of its electricity, and if, after this has been done, we present a piece of *sealing-wax* in place of the glass formerly employed, the very same phenomena will take place. On the *first* application the ball will be *attracted*, and on the *second repelled*. It is clear then, in the first place, that both these electrics have the power of attracting another body, before they have communicated to it any of their own electricity; and secondly, that they repel the body, after they have communicated to it a portion of their own electricity.

6. But a very remarkable circumstance takes place if we, after having conveyed electricity to the ball B, by means of excited glass, which was for a moment or two in contact with it, should present to it after the former had been withdrawn, *excited sealing-wax*, the ball, instead of being *repelled*, as it would have been if the glass had been applied, is *attracted by the wax*. If the experiment be reversed, and the excited wax first presented to the ball, and then the excited glass, the *latter* will be found to *repel* the ball. Hence it follows, says Sir David Brewster, that excited glass *repels* a ball electrified by excited glass. Excited *wax* repels a ball electrified by excited wax. Excited *glass attracts* a ball electrified by excited *wax*. Excited *wax attracts* a ball electrified by excited *glass*. From this we conclude, that there are two opposite electricities, namely, that produced by excited glass, called *vitreous* or *positive* electricity; and that produced by excited wax, called *resinous* or *negative* electricity; such electricity being obtained by *friction*. But as nature always works by the simplest means, it seems more consistent with her usual operations, that there should be one fluid rather than two, provided that known facts can be equally well accounted for

7. Opposite electricities always accompany each other, for if any body become positive, the body with which it is rubbed becomes negative. Where two bodies oppositely electrified are united, their powers are destroyed, and if the union be made through the human body, it produces an affection of the nerves called an *electric shock*.

THE ELECTRICAL MACHINE.

8. Machines have been contrived for rubbing together the surfaces of electrics and non-electrics, and for collecting the electric fluid when so excited. They are made both in the

form of a cylinder, and of a circular plate. The latter is represented at the head of the chapter. The plate is turned by the handle through the rubber, which is coated with a metallic amalgam, and diffuses the excitement over the glass. The points carry off a constant stream of positive electricity to the prime conductor. Negative electricity is obtained by insulating the conductor to which the cushion is attached, and connecting the prime conductor with the ground, so as to carry off the fluid collected from the plate. If the person who works the machine be supported on a stool having glass legs, and connected with the conductor by means of a metallic rod, or if he touch it with his hand, he is found to be in the same state of electricity, and another person standing upon the ground, can draw sparks from him by presenting his knuckles to his body; for the glass prevents the electric fluid from running from him to the earth, and therefore having more than his share of it, he will be ready to part with it to any surrounding bodies.

9. The process of the electrical machine is this:—By means of the chain trailing on the ground, the electric fluid is collected from the earth in the glass cylinder, which gives it, through the points, to the conductor. A person may act the part of the chain in conducting the fluid from the earth to the cushion, while the plate or cylinder is turned, and the machine will work equally well. The principle, therefore, of the electrical machine is by excitation to collect the fluid, which cannot escape again, owing to the glass cylinder globe or plate being insulated.

10. If, instead of acting on the chain, a person places himself on a stool with glass legs, all communication will be cut off between the cushion and the earth; in other words, the cushion will be insulated, and the machine will only take from him what it can get from his body. In this case, the person will have *less* than his natural share of electricity, and will be electrified *minus* or negatively; and if another person comes near to his body, he will part with a portion of his electricity to make up for the other's loss; or rather the second person's body will act as a conductor to convey the electricity from the earth.

11. Suppose A stands on the glass legged stool and holds the rubber, and B stands on another glass legged stool and touches

the prime conductor while the machine is turned. It is evident that A will be *negatively* electrified and B *positively*; because A will lose part of his electricity, which will be given to B, and thus A will be minus or less than his natural share, and B plus or more than his natural share. If a third person, C, be placed on another glass-legged stool, he, by touching B, will take from him what he has too much, and by coming in contact with A, he will convey to A again what he had lost.

12. THEORIES OF ELECTRICITY.—The hypothesis which would account for electrical phenomena, assumes that there is an extremely subtile and highly elastic fluid pervading all natural bodies, and capable of moving with various degrees of facility through the pores or actual substance of various kinds of matter. In some, as in those we call *conductors* or *non-electrics*, it moves *readily*; and with others, called *non-conductors* or *non-electrics*, it moves with difficulty. But there is another theory, which supposes this fluid to be *a compound*, susceptible of decomposition by friction and other means, and hence the origin of the term *vitreous* and *resinous* electricity.

13. EXCITATION.—From various causes, of which the friction of surface is one, the state of union in which two electrics exist in bodies is disturbed: the vitreous electricity is impelled in one direction, while the resinous is transferred to another, and each manifests its peculiar powers. When accumulated in any quantity, each fluid acts in proportion to its relative quantity; that is, to the quantity which is in excess above that which is still retained in a state of neutrality by its union with electricity of an opposite kind. Hence, when two bodies are rubbed together, say glass and a metallic amalgam only, the electricity at the surfaces subjected to such friction is decomposed, the *resinous* adhering to the *amalgam*, while the *vitreous* attaches closely to the *glass*. All the rest of the electricity remaining on each surface undecomposed, is in a state of perfect quiescence or inertness.

14. DISTRIBUTION.—Both of these fluids being highly elastic, their particles repel one another with a force which increases in proportion as their distance is less; and this force acts at all distances, and is not impeded by the interposition of bodies of any kind, provided they are not themselves in an active electrical state. It has been deduced from the most careful analysis, that this force follows the same law as gravita

tion, viz., that its intensity is inversely as the square of the distance.

15. TRANSFERENCE.—Since the two electrics have this powerful attraction for each other, they would always flow towards one another and coalesce, were it not for the obstacles thrown in their way by *non-conductors*. When, instead of these conductors, fluid substances are interposed, they enter into union with great velocity, producing, in their transit and confluence, several remarkable effects. When once united, these powers remain dormant until again called into action by the renewed separation of the fluids.

16. ATTRACTION AND REPULSION.—The repulsion which is observed to take place between bodies that are insulated and charged with any one species of electricity, and other bodies similarly charged, is derived from the repulsive power which the particles of this fluid exerts towards those of their own species; and the attractions between bodies differently electrified is derived from the attractive power of the vitreous particles for those of the opposite kind.

17. Thus far we have proceeded upon the hypothesis of *two* distinct *fluids*. But the Franklin theory supposes that one, which we call the *electric fluid*, is highly elastic, or repulsive of its own particles; and there is hardly any fact that is explained on the hypothesis of two fluids which is not equally explicable on that of one; and the explanations of the first are easily converted to those of the second, by substituting the expressions positive and negative for those of vitreous or resinous electricities. The principal advantage of Franklin's theory is its extreme simplicity.

18. INDUCTION.—If a body is charged with electricity, and insulated so perfectly as to prevent the escape of the electricity which it contains, it nevertheless tends to produce an electrical state of the *opposite* kind in all the bodies that surround it. Thus, the vitreous induces the resinous, or the resinous the vitreous electricity, in a body that is situated in the vicinity of either of them; and this to a degree proportionate to the smallness of the distance that separates the bodies. The electricity in this case is said to be *induced*, and the phenomenon is called *electrical induction*.

19. The most convenient mode of obtaining an accumula

tion of electricity arising from induction, is by the employment of a plate of glass, on each side of which is pasted a sheet or coating of tin foil. The form of coated glass best adapted to experiments is that of a cylindrical jar, coated within and without nearly to the top. The cover consists of baked wood, and is encased with sealing wax, to exclude moisture and dirt. A metallic rod, rising two or three inches above the jar, and terminating at the top in a brass knob, is made to descend through the cover till it touches the interior coating.

20. The LEYDEN JAR is so called, in consequence of the discovery respecting the effects of the union of two bodies, apparently electrified, being made at Leyden, in Holland, by means of a phial or small bottle. It is used in the following manner: — The outer coating being made to communicate with the ground, by holding it in the hand, the knob of the jar is presented to the prime conductor when the machine is in motion. A succession of sparks will pass between them, while at the same time nearly an equal quantity of electricity will be passing out from the exterior coating, through the body of the person who holds it to the ground.

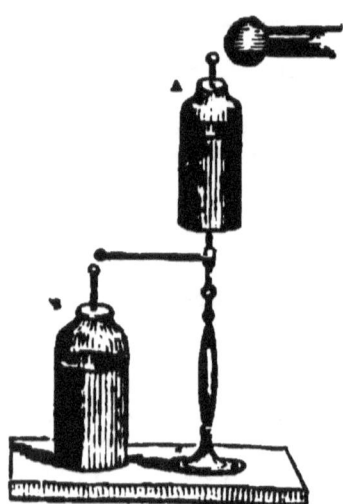

The jar, on being removed, is said to be charged; and if a communication is made between the two coatings, by a metallic wire extending from the external one to the knob, the electric fluid which was accumulated in the positive coating passes along the conductor into the negative coating with a loud snap and flash of light, thus at once restoring an almost complete equilibrium. In the cut two jars are shown. Here, for every spark which passes from the conductor to the inside of A, a corresponding spark will pass from the outside of A to the inside of B, until both jars are charged.

21. By uniting together a sufficient number of jars, we are

able to accumulate an enormous quantity of electricity. For this purpose, all the *interior* coatings of the jars must be made to communicate by metallic rods, and a similar union must be established among the *exterior* coatings. When thus arranged, the whole series may be charged as if they formed but one jar, and the whole of the accumulated electricity may be transferred from one system of 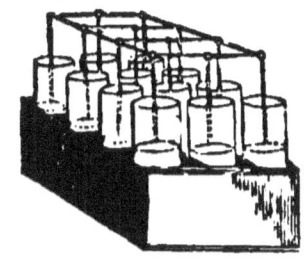 coatings to the other by a general and simultaneous discharge. Such a combination of jars is called an Electrical Battery.

22. For the purpose of making a direct communication between the inner and the outer coatings of a jar or battery, by which a discharge is effected, an instrument called a discharging rod is employed. It consists of two bent metallic rods, terminating at one end by brass balls, and, connected at another by a joint, which is fixed at the end of a *glass* handle, and which, acting like a pair of compasses, allows of the balls being  separated at different distances. When opened to the proper degree, one of the balls is made to touch the exterior coating, and the other ball is then quickly brought into contact with the knob of the jar, when a discharge is effected, while the glass handle secures the person holding it from the effects of the shock.

23. ELECTROMETER.—The electrometer is an instrument used for the purpose of detecting the presence of electricity. The engraving represents one which depends entirely upon the repulsion which takes place between two bodies in a state of electrification, terminating in a knob, to prevent the electricity from passing away. It consists of a light rod and a pith ball, hanging parallel to the stem; but turning on the centre of a graduated semi-circular scale. The machine may be

taken off the stand, and is to be placed in a hole made for it in the conductor, and according as this is more or less electrified, the ball will fly further from the stem, and the degrees of the scale will exhibit the intensity of any given charge. Two pith balls may also be suspended parallel to one another on silken threads, and applied to any part of an electrical machine, and they will, by their repulsion, serve for an electrometer. This arrangement will also show whether the electricity be positive or negative; for, if it be positive, by applying an excited stick of sealing-wax, the threads will go together again; but if it be negative, then, excited sealing-wax, or resin, or sulphur, or even a rod of glass, the polish of which is taken off, will make them recede further.

24. The UNIVERSAL DISCHARGER.—This instrument is used for passing charges through various substances which may be laid on the plate, D. B B are glass pillars cemented to the frame A, and to each of these is cemented a brass cap with a double joint for horizontal and vertical motion; on the top of each joint is a spring tube, which holds the sliding wires C x, C x, so that they may be set at various distances from each other, and turned in any direction. The extremities of the wires are pointed, but with screws at about half a inch from the points, to receive balls. The rings C C receive the chain or wire from the conductor.

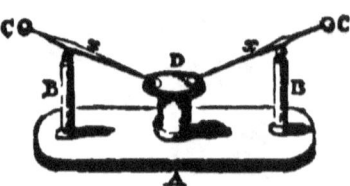

25. DE LUC'S COLUMN.—This instrument consists of a glass tube (A) closed at each end by a brass knob, and containing a number of pieces of Dutch leaf, with paper placed between them. By this means electricity is produced, and may be made to attract the pith ball, B, which is suspended by a silk thread between the column and C, a piece of wood covered with tin foil communicating by a chain with the ground. After the ball has become charged, by contact with the column, it is re-

pelled, and then flies to the tin foil conductor, where it parts with an excess of electricity, and becomes *negative*, returning to its perpedicular position, to be again attracted and repelled as before. By this means the ball will continue in action as long as any electricity is generated, and this is a very near approach to *perpetual motion*.

26. WORKING POWER OF ELECTRICITY.—This may be shown by the subjoined machine.
A is a wooden board, B B B B, four pillars, having fine wires (c c) stretched above. On these rests the rotatory wire or wheel EE, having its points turned reversed ways. By means of a chain attached to the conductor, and to the in-

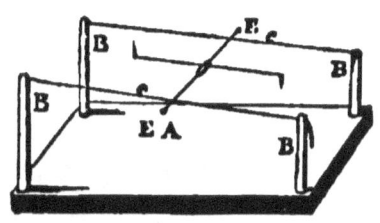

strument at B, the electricity passes over the pillar B, up the wire c unto the wheel, and off at the points, and causes it to turn round upon an inclined plane, till it reaches the top.

27. RINGING BELLS.—The engraving represents three bells suspended from a brass wire, D D, and supported by a glass pillar, A, passing through the bell B to the bell E. The electrical apparatus being attached to the knob at E, the electricity passes down the wires D D to the bells, which are then *positively* electrified, and attract the clappers C C, that are *negatively* so, in consequence of being insulated by the silken

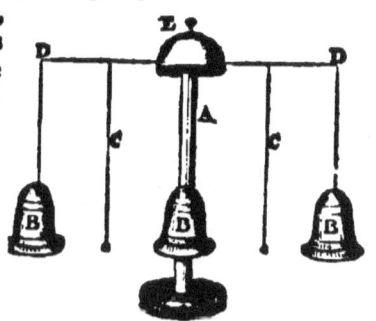

strings, which are non-conductors. The bells therefore attract the clappers, till they are charged, when they strike against the centre bell to discharge themselves, and thus a peal is rung on the bells until the electricity is taken off.

28. OF THE ELECTRIC SHOCK.—If a person who is standing receives an electric shock through the spine, he loses the power of the muscles to such a degree, that he drops motionless. A strong charge passed through the head, gives the sensation of

a violent, but universal blow, and is followed by a transient loss of memory and an indistinctness of vision. If the diaphragm be included in the current of a coated surface of two feet in extent, fully charged, the sudden contraction of the muscles of respiration will act so violently upon the air in the lungs, as to occasion a loud and involuntary shout; but if the charge be small, a fit of convulsive laughter is induced, producing a most ludicrous scene to the bystander. Small animals, such as mice or sparrows, are instantly killed by a shock from 30 square inches of glass.

29. ELECTRIC SPARK.—The spark is of a bluish colour in the atmosphere in its ordinary state; but in a glass vessel

containing condensed air, it is white; in rarefied air it is of a redish tinge; and in a good vacuum, under a receiver, it is of a purple hue very faintly visible. When the spark is very long, it presents the appearance represented in the cut. If a person on the

electral stool be charged with much electricity, his hair will stand on end "like quills upon the fretful porcupine." A wooden head, in honour of the philosophers, is prepared with some long hair on it; this if placed by a wire on the conductor, and the machine be put in action will exhibit itself as in the engraving.

30. IDENTITY OF LIGHTNING WITH ELECTRICITY.—To demonstrate the identity of the electric fluid with lightning, Dr. Franklin contrived to bring the lightning from the heavens in the following manner:—Having constructed a kite by stretching a large silk handkerchief over two sticks in the form of a cross; on the appearance of an approaching storm, he went into a field in the vicinity of Philadelphia, and raised it in the air, taking care to insulate himself, by a silken cord attached to a key, with which the hempen string

terminated. No sooner had a dense cloud, apparently charged with lightning, passed over the spot on which he stook, then his attention was arrested by the bristling up of some loose fibres on the hempen string; he immediately presented his knuckle to the key, and received the electric spark. From this key he charged the phials, and from electric fire thus obained, he performed all the common electrical experiments. This discovery has led to the preservation of buildings from the effect of lightning, by passing a rod of metal from their highest points into the ground, by which the electric fluid frequently travels without injury to the building.

31. ANIMAL ELECTRICITY.—The torpedo, the electrical eel, and some other fishes of the ray genius, communicate shocks, upon being touched with the hand, or by electrical conductors. The membranous organs, which produce this effect, are like the cells of a galvanic trough, or of a bee-hive, and are very distinctly marked. The shock of the torpedo seldom extends above the touching finger, and never above the elbow; but it can give 20 in a minute. These animals seem to use the electrical property as a means of self-defence.

32. ELECTRICITY AS EVOLVED BY THE CONTACT OF DIFFERENT METALS.—Thus, if two discs, the one of copper and the other of zinc, rather more than two inches in diameter, and furnished with insulating handles, be brought into contact, and then separated and examined by the electroscope, the copper disc is found to be charged with the *negative*, and the zinc disc with the *positive* electricity. While the contact of the metals is preserved, neither of them gives any indication of its electrical state. This relates to that mode of electrical excitement called *galvanism*.

33. Some bodies are rendered electrical by pressure: thus, if a crystal of calcareous spar be pressed for a few minutes between the fingers, it exhibits *attraction*. The same may happen with regard to cork, paper, and wood. Many mineral substances, when reduced to powder, exhibit electricity if made to fall on an insulated metallic plate, a mode of excitation which is to be considered as a species of friction.

34. There are several mineral bodies, which, from being in a neutral state at ordinary temperatures, acquire electricity simply by being heated or cooled. This property is confined

to crystalized minerals; of these the most remarkable are tourmaline and boracete. In the former of these, it is best observed in the regularly terminated crystals. When one of these is heated from 100° to 212° Fah., the extremity, terminated by the greatest number of planes, becomes charged with *positive* electricity, while the other extremity is *negative*.

35. A large number of substances become electrified in passing from the liquid to the solid form. This happens to sulphur, gum lac, bees-wax, and in general all resinous bodies. The conversion of bodies into a state of vapour, as well as the condensation of vapour, is generally attended by some alteration of their electrical condition. Thus, if a red hot platina crucible be placed upon the gold leaf electrometer, and water be dropped upon it, at the moment the vapour arise, the leaves of the electrometer diverge with negative electricity.

36. DECOMPOSITION AND REPRODUCTION OF WATER BY ELECTRICITY.—When a succession of electrical discharges from a powerful electrical machine are sent through water, a decomposition of that fluid takes place, and it is resolved into its two elements oxygen and hydrogen, which immediately assume the gaseous form. When this experiment is conducted in a suitable apparatus, and a shock is transmitted through the mixed gases thus obtained, they are instantly kindled, and a re-union of the elements takes place, and precisely the same quantity of water is reproduced as was decomposed to furnish the gases.

37. PHENOMENON OF ELECTRICITY IN A THUNDER STORM. —In a thunder storm the clouds are non-electrics or conducting surfaces, positive, with a negative sphere extending to the earth; and the discharge at a point from one large surface to the other is the lightning; or the earth is negative, and the clouds correlatively positive. All bodies in the sphere of action are affected. The cloud, the air, and the earth, resemble the zinc, fluid, and copper, in a galvanic combination, which will be explained in the subsequent chapter.

GALVANISM.

CHAPTER LII.

DISCOVERY BY SULZER—GALVANI—MODE OF EXCITATION—GALVANIC PILE—TROUGH BATTERY—CHEMICAL, ELECTRICAL, AND PHYSIOLOGICAL EFFECTS.

1. GALVANISM is now identified with electricity, and may be considered a branch of that science. The first notice we find of this peculiar form of electric agency occurs in a metaphysical work published in 1767, by a German writer named Sulzer, who observed that by applying two metals, one above, and the other under the tongue, and then bringing them in contact, a peculiar taste was perceived. Sulzer did not trouble himself to ascertain the true cause of this phenomenon, which he hastily attributed to a vibratory motion excited by the contact of the metals, and communicated to the nerves of the tongue; but about the year 1791, Galvani,[*] Professor of Anatomy at Bologna, made some important discoveries on this subject. He found, that independent of the usual modes of exciting electricity upon animal substances, that similar effects might be produced by the development of electricity at the surface of contact of different metals.

2. THE CHEMICAL THEORY OF GALVANISM is that chemical action occurring between a fluid and a solid body, is always accompanied by the disturbance of electric equilibrium, and thus a quantity of electricity passes from a latent into an active state.

3. The process usually adopted for producing galvanic elec-

[*] The science of galvanism is said to have arisen from the following circumstance:—A recently killed frog having been accidentally touched in the limb with the blade of a knife which was held by a person who was experimenting with an electrical machine, was immediately thrown into violent convulsions. Galvani was not present when this occurred, but being informed of the circumstance, he lost no time in repeating the experiment; and extending his observations upon the phenomenon, he found that other metals besides that forming a knife, answered the purpose, and very justly inferred that they owed this property of exciting muscular contractions to their being good conductors of electricity.

tricity, is to interpose between two plates of different kinds of metal, a fluid capable of exerting some chemical action on one of the plates, while it has a different one upon the other, and then to establish a communication between them, which constitutes what is called the galvanic circle. When these plates and chemical substances are multiplied, galvanism is more forcibly exhibited, increasing in proportion with an increased number of plates. To this multiplication of plates belongs a machine called the *galvanic pile*, discovered by Volta, and hence called the *voltaic pile*.

4. THE VOLTAIC PILE is composed of alternate layers of

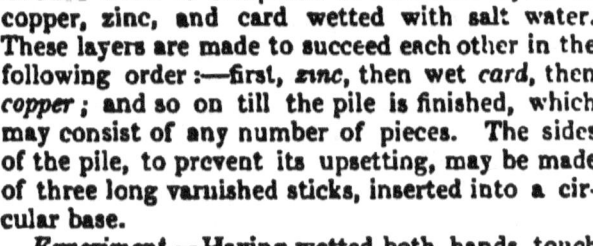

copper, zinc, and card wetted with salt water. These layers are made to succeed each other in the following order :—first, *zinc*, then wet *card*, then *copper*; and so on till the pile is finished, which may consist of any number of pieces. The sides of the pile, to prevent its upsetting, may be made of three long varnished sticks, inserted into a circular base.

Experiment.—Having wetted both hands, touch the lower parts of the pile with a piece of copper wire, and the upper part with another piece, and smart shocks will be felt, increasing with the chemical action of the plates upon each other.

5. GALVANIC TROUGH.—Various modes have been adopted of increasing the galvanic action, and what is called a *trough* battery has been invented. This consists of a long trough made of non-

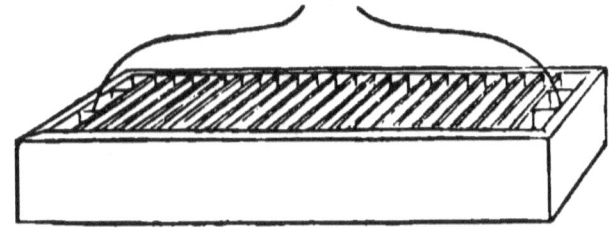

conducting materials, and divided into compartments. Into this a fluid (diluted nitric acid) is poured, into which is put a series of plates of zinc and copper, united into pairs, a pair of each falling into every compartment of the trough. A number of these troughs may be united together, by connecting together the terminal plates of the adjoining troughs by slips of copper, taking care to preserve the order of the series. The voltaic battery belonging to the Royal Institution is constructed upon this plan, and consists of 200 separate parts, each composed of ten double plates, and each plate containing 32 square inches, the whole number of double pieces being 2,000, and the whole surface 128,000 square inches. The mode in which the trough or battery is made to act, is by bringing the ends of the wires together, on the plate of glass represented below. If a grain of gunpowder be laid between the points of the wires it will explode; and when any substance is brought within the shocks of the great battery alluded to above, it instantly becomes ignited.

6. IMPROVEMENT OF THE VOLTAIC BATTERY.—The voltaic battery has been improved by keeping the plates detached, instead of soldering them together. They are connected at the upper edge by a metallic arc, and are introduced into a trough divided into cells by partitions of

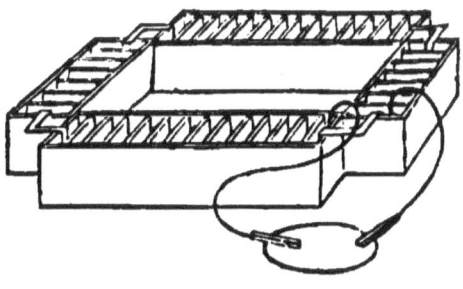

glass (or sometimes into troughs wholly made of earthenware), in such a manner that on one plate is one side of the partition, the other on the other. This arrangement has the advantage, that, both surfaces of each plate being acted on, a greater power is obtained.

In the engraving the plates are shown suspended over a porcelain trough, which is the best form in which they can be constructed. The only practical objection to the arrangement is, that in some cases, the acid

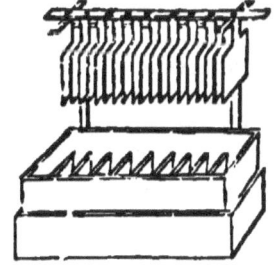

acts on the glazed surface of the porcelain, and the troughs leak

7. Dr. Wollaston has heightened the improvement, by placing in each cell one plate of the one metal, as the zinc, and two of the other, the copper, so that each surface of the zinc may be opposed to a surface of copper. The plates of copper are connected by metallic arcs, both at the top and bottom: and between them, supported by pieces of wood, is the plate of zinc, distant an eighth or fourth of an inch from the copper on each side. The communication between these triple plates is established by arcs of lead or other metal, connecting each central zinc plate with the copper of the adjoining cell. This arrangement is very powerful in producing light and heat. A single series of this description is shown in the engraving.

8. GALVANIC DEFLAGRATOR.—An ingenious modification of this apparatus has been contrived by Dr. Hare, of Philadelphia. It consists of concentric coils of copper and zinc, so suspended by beams and levers as to be made to descend, at pleasure, instantaneously into the exciting fluid contained in glass jars or wooden troughs, without partitions. Each coil is formed from a zinc sheet of 9 inches by 6, and one of copper 14 inches by 6, more of the copper being required, as this metal is made to commence within the zinc, and completely to surround it without. 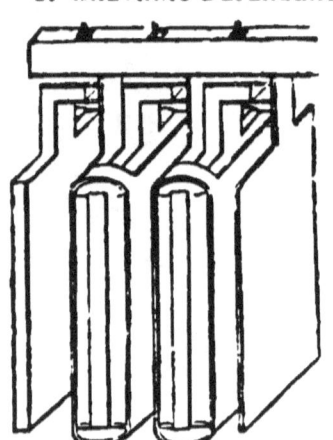 The sheets are so coiled as to leave between them interstices of a quarter of an inch. In the original apparatus, they were arranged in two rows, 40 coils in each: on their immersion in the appropriate fluid, the immediate evolution of heat and light was found to be most intense, and far exceeding that of voltaic piles or troughs of an equal number of series and extent of surface; and on account of its superior power of causing the combustion of metallic wires and leaves, the instrument was named by its inventor the *galvanic deflagrator*.

9. ELECTRICAL EFFECTS OF GALVANISM.—Galvanism, even when excited by a single galvanic circle, distinctly affects the

electrometer. If the ball of the electrometer touch the copper the electricity is negative. When wires connected with the opposite poles or sides of an active galvanic trough are brought near each other, a spark is seen to pass between them, accompanied by a slight snap or a report. Upon establishing a communication by means of hands, an electrical shock is propagated through a number of persons, without any perceptible interval of time.

10. CHEMICAL EFFECTS OF GALVANISM.—By the galvanic battery is performed the ignition and fusion of metals. The facility of ignition in the different metals, appears to be conversely proportionate to their power of conducting heat. Hence platina, which has the lowest conducting power, is most easily ignited; and silver, which conducts heat with greater facility than any other metal, is ignited with more difficulty than any of the rest. The metals are burnt, or rather deflagrated, in the form of very thin leaves. Gold emits a very vivid white light, inclining to blue. The flame of silver is a vivid green, somewhat like that of the emerald; and zinc gives forth a bluish white flame, fringed with red.

11. By galvanic batteries the most surprising effects have been produced. Electric fire of the most extraordinary intensity is evolved from wires proceeding from the different poles of a battery, when brought within a short distance of actual contact. In the galvanic flame thus produced the most refractory substances, as quartz, the sapphire, fragments of diamonds, and points of charcoal and plumbago, have been fused, and many substances, formerly supposed to be simple, have been decomposed into their constituent parts. It was by this power that Sir Humphrey Davy discovered the metallic bases of the alkaline earths already explained.

12. PHYSIOLOGICAL EFFECTS OF GALVANISM.—The Physiological effects of galvanism are very striking. The action of this kind of electricity on a dead frog, as well as on other animals, occasions a tremulous motion of the muscles, and generally an extension of the limbs. If the legs of a frog recently dead be skinned, and a small part of the spine be attached to them, but separate from the rest of the body, and a part of the nerve proceeding from this limb be wrapped up in a bit of tin foil, or laid upon zinc, the motion of the limb will be very vigorous; the two metals may be placed, not in contact with

the preparation, but in any other part of the circuit, which may be completed by means of other conductors, as water, &c.

13. The impression made upon some of the nerves of the face when they form part of the circuit, is accompanied by the sensation of a vivid flash of light. When a piece of zinc and a piece of copper are placed, the one above and the other below the tongue, which must be in a moist state, a peculiar taste is experienced. This is supposed to arise from the saliva of the mouth having been decomposed by the galvanic action, and not merely the effect of a direct impression of the electric current on the nerves of the tongue. When the current of voltaic electricity is made to pass along a nerve distributed to any of the muscles of voluntary motion, they are thrown into violent convulsive contractions. The susceptibility of some animals is very great, and numerous curious experiments may be performed with them. If an earthworm be placed upon a crown piece which lies upon a plate of zinc of larger size, it will suffer no inconvenience as long as it remains in contact with the silver only; but the moment it has stretched out its head, and touched the zinc, so as to complete the galvanic circle, it suddenly recoils, as if it had felt a severe shock.

14. When large animals are made the subject of experiment, the effects are still more striking. Thus, if two wires, connected with the poles of a battery of 100 plates, be inserted into the ears of an ox or sheep, when the head is removed from the body of the animal recently killed, very strong action will be excited in the muscles of the face every time the circuit is completed. The convulsions are so general, as often to impress the spectator with a belief that the animal has been restored to the power of sensation, and that he is enduring the most intense sufferings; the eyes are seen to open and shut spontaneously, they roll in their sockets, as if again endowed with vision, the pupils being at the same time widely dilated, the nostrils vibrate as in the act of smelling, and the masticatory movements of the jaws are imitated. The effects of some experiments performed upon the human subject, a malefactor executed at Newgate, were so awful as actually to frighten one or more of the students who witnessed them; the body raised itself suddenly up, lifted its arm, and with the eyes and the muscles of the face depicting the most intense agony, struggled apparently in convulsions.

MAGNETISM.

CHAPTER LIII.

THEORY—NATURAL MAGNET—POLARITY—ARTIFICIAL MAG-
NET—MARINER'S COMPASS—AZIMUTH COMPASS—DIP OF
THE NEEDLE—VARIATION—ANIMAL MAGNETISM—ELEC-
TRO-MAGNETISM—THERMO-ELECTRICITY.

1. The THEORY OF MAGNETISM bears a very strong resemblance to that of electricity. Like it, magnetism has its attractions and repulsions, and it can be excited in one body and transferred to another; with, however, this striking peculiarity, that carbonised iron is very nearly the only substance capable of exhibiting any strong indication of its presence.

2. The NATURAL MAGNET has been known from a very remote period. It appears native in a grey iron ore in octahedron crystals, composed of from 73 to 85 of iron, and from 15 to 25 of oxygen. It acquires its magnetic power after exposure to the air, displays polarity if freely suspended, one end pointing to the north and the other to the south. It attracts iron, and will lift from 40 to 50 times its own weight; its force or power diminishing in the inverse of the distance. The natural magnet is, however, seldom used, as its properties can be imparted to bars of steel which may be made far more powerful than itself.

3. POLARITY.—When a bar of steel thus magnetised is freely suspended, as it may be on a cork in a basin of water; like the natural magnet one of its ends will point towards the north and the other towards the south. This is termed *polarity*, and that end which points to the north is called the north pole, and that which points to the south the south pole of the magnet. If the north pole of a magnetic bar be presented to a similar pole of another, it will be repelled; but if two oppo-

site poles be presented, they will be attracted, the same as negative and positive electricity.

4. MARINER's COMPASS.—An artificial magnet, fitted in a proper box, for the purpose of guiding the traveller, is called a magnetic needle, and the whole together is called the mariner's compass. This instrument consists of three parts:—1. The box; 2. The card or fly; 3. The needle. The box, which contains the card with the needle, is made of circular form, and either of wood, brass, or copper. It is suspended within a square wooden box, by means of two concentric brass circles called *gimbalds*, so fixed by cross axes to the two boxes, that the inner one, or compass-box, shall retain a horizonal position in all motions of the ship, whilst the outer or square box is fixed with respect to the ship. The compass-box is covered with a pane of glass, in order that the motion of the card may not be disturbed by the wind.

5. What is called the card, is a circular piece of paper, which is fastened upon the needle, and moves with it. Sometimes there is slender rim of brass, which is fastened to the extremities of the needle, and serves to keep the card stretched. The outer edge of this card is divided into 360 equal parts or degrees, and within the circle of those divisions it is again divided into 32 equal parts or arcs, which are called the *points of the compass*, or *rhumbs*, each of which is often subdivided into quarters. The initial letters N, N E, &c. are annexed to those rhumbs, to denote the north, north-east, &c. The middlemost part of the card is generally painted with a star, whose rays terminate in the above-mentioned divisions.

6. The magnetic needle is a slender bar of hardened steel, having a pretty large hole in the middle, to which a conical piece of agate case is adapted, by means of a brass piece, into which the agate is fastened. The apex of this hollow cap rests upon the point of a pin, which is fixed in the centre of the box, and upon which the needle, being properly balanced, turns very nimbly. For common purposes, these needles have a conical perforation made in the steel itself, or in a piece of brass which is fastened in the middle of the needle.

7. The needle, which is balanced before it is magnetized, generally loses its balance by being magnetized, on account of the dipping: a small weight, or moveable piece of brass, there-

fore, is placed on one side of the needle, by the shifting of which, either nearer to, or farther from the centre, the needle will always be balanced.

8. THE AZIMUTH COMPASS is nothing more than the above-mentioned compass, to which two sights are adapted, through which the sun is to be seen, in order to find its azimuth, and from thence to ascertain the declination of the magnetic needle at the place of observation; in one of these there is an oblong aperture, with a perpendicular thread or wire stretched through its middle, and in the other sight there is a narrow perpendicular slit.

9. VARIATION OF THE NEEDLE.—The magnetic needle does not point exactly north and south; but the north pole of the needle takes a direction considerably to the west of the true north. This is called the variation of the needle, the cause of which is not understood. It is constantly changing, and varies at different parts of the globe, and at different periods of the day and night. In 1657, the needle pointed due north, previously its variation had been to the east, but it from that period took a westerly direction. In the year 1814 it attained its greatest variation, and from that time to the present it has been slowly returning, and in about 120 years it will perhaps again reach the limits of its deviation.

10. THE DAILY VARIATIONS of the magnet are from 8 o'clock, A.M., when the declination increases, till about 3 o'clock, P.M., from which time it remains unaltered until 8, A.M. The amount of these deviations is the greatest from April to July, when it is 13' to 16', in the other months it is from 8' to 10'. The direction of the needle is affected by approaching earthquakes or eruptions of volcanoes. If a needle stands on the magnetic meridian, and is displaced by a foreign power, it returns, when the power ceases to act, to its former situation, by a series of *oscillations.* These determine the intensity of magnetism. The time of an oscillation in the case of the same needle, has also a certain relation to the magnetic power of the earth, and serves as a measure of it in a similar way as the oscillations of the pendulum serve for the measurement of the degrees of gravity.

11. DIP OF THE NEEDLE.—Another remarkable and evident manifestation of the influence of the magnetism of the earth upon the needle, is the inclination or dip of the latter, which

is a deviation from its horizontal plane in a downward direction, in northern regions of the North Pole of the magnet, and in southern regions of the South Pole; and which, in the regions of the magnetic equator is 0, but increases towards the poles. This inclination in the northern hemisphere is greatest between 70° and 80° latitude; under 74° 47', where Parry remained during the winter the dip or inclination amounted to 88° 43'' 45'. The cause of all these phenomena is yet unexplained.

12. ANIMAL MAGNETISM.—This fanciful science appears to have originated with Mesmer and other German illuminés, who believed that the power of a common magnet might be made to act on the human frame. They also believed that this power once imparted to a professor, might be communicated to others without the agency of the magnet. The phenomena induced are faintings, convulsions, spasms, lively dreams, insensibility to pain, fits, &c. The magnetised person, after being blindfolded, is *said* to read letters from the pit of his stomach or with his toes. To succeed fully in these experiments, it is indispensible that both the subject and the witnesses should be of a disposition to believe without doubting, and sceptics are consequently never enlightened.

ELECTRO-MAGNETISM.

13. ELECTRO-MAGNETISM.—This is the name applied to a very interesting class of facts, principally developed by Professor Faraday, who, in a series of very curious experiments, has succeeded in identifying magnetism and galvanism, by directing galvanic currents at right-angles to the position of powerful magnets, and has produced the galvanic currents from terrestrial magnetism, and thus proved that the same causes or elementary disturbances produce the directive character of both.

14. A current of galvanic action changes the position of the magnetic needle from north and south to east and west. This is explained on the principle, that as each pole had its own determined electricity, so a current of common electrical action would accord with neither, and the consequent re-action would place the two poles at a right-angle to the current. Polarity is therefore inferred to be the effect of an electrical current

following the heat of the sun, and directing the natural arrangement of the loadstone into a direction of right-angles, or towards the poles.

15. The same influence which affects the magnetic needle, also communicates magnetism to soft iron. If a bar of that metal be surrounded with a copper wire, prevented from touching the iron by a winding of cotton or thread, as a common bonnet wire, and then if a current of voltaic electricity be sent through the bar, it becomes a powerful magnet and will continue so, as long as connection with the battery is preserved. On breaking the contact, the magnetism disappears. This experiment is illustrated by the annexed figure, which is a horse-shoe magnet, surrounded by several coils of wire. P is the positive and N the negative pole.

16. When an iron armature is made to revolve in *approximation* with the steel magnet, the former being invested in a coil of copper wire, in the same manner as an electro magnet, the iron becomes highly electrical by induction, and exhibits all the properties common to a voltaic battery. The following experimental arrangement exhibits this fact, and fully illustrates the identity of magnetism with electricity and galvanism. It consists of a compound magnet in the horse-shoe form, resting against a vertical mahogany board, behind which is a wheel, which, by means of a band, communicates rapid motion to the armature, and the two coils of copper wire. By this motion the inherent electricity of the wire coils is called into action by the disturbing influence, and by the stream of electricity evolved from the poles of the magnetic battery acting on the iron cylinders evolved by the wire so that each time the armature, P, during its

rotation, is vertical with its iron cylinder between the poles of the magnet, a brilliant spark is evolved from the point *a*. By this a variety of beautiful experiments may be performed, such as the ignition of metals, firing of gunpowder, and the decomposition of water, while shocks similar to those of electricity and galvanism, may be taken.

17. MAGNETS MADE BY GALVANISM.—To effect this, make a connection between the poles of an excited battery, with the two ends of a wire formed into a spiral coil, round a cylinder or tube of about an inch in diameter. Into this coil introduce a needle or piece of steel wire, laying it lengthways, down the circles of the coil. In a few minutes after the electric fluid has passed through the spiral wire, the latter will be found to be strongly magnetized, and to possess all the properties of a strong magnet.

18. EFFECT OF A CURRENT OF ELECTRICITY UPON A MAGNET.—If a current of electricity be made to pass along a wire, under which, in a line with it, a compass is placed; it will be found that the needle will no longer point north and south, but take a direction nearly across the current, and point almost east and west—from which it is inferred, that constant currents of electricity are passing in that direction across the earth.

19. THERMO-ELECTRICITY is that electricity developed under the influence of *temperature*; and was first observed and made known by M. Seebeck, in the year 1822. Experiments have shown that the thermo-electric properties of metals have no connection with their galvanic relations, or their capacity for conducting heat or electricity; neither do they accord with their specific gravity or their atomic weights.—If a piece of platina and a similar piece of iron wire are slightly twisted together, and their ends immersed in the mercury cups of the galvanometer, whilst their opposite ends are held in the flame of a spirit lamp, the needle will be deflected : this is an instance of thermo-electricity.

Illustrations.
MOTION AND MECHANICS.

CHAPTER I.

ATTRACTION OF MOUNTAINS—EXPERIMENTS OF CAVENDISH—DISSEC-
TION OF CRYSTALS—GIANT'S CAUSEWAY—BALANCING—MOMEN-
TUM—CENTRE OF GRAVITY—ANIMAL MECHANICS, ETC

ILLUSTRATION I.

ATTRACTION OF MOUNTAINS.—The greatest mountain on the earth s surface is not the fifty-nine millionth part of its bulk. The attraction of such a mountain, therefore, upon a ball of lead, is as nothing, compared with the attraction of the whole earth upon a ball of lead; yet this attraction produces a sensible deviation of the plumb-line. On the sides of Chimboraço, the highest of the Andes, by observations made with great difficulty, Bouguer ascertained that mountain to attract the plumb-line 7 or 8 seconds from the perpendicular. In 1772 Maskelyne, by similar observations made at the foot of Mount Sbehallian, in Scotland, found a deviation of the plumb-line of 54 seconds. These experiments enable us to compute the mean density of the earth; for, by observing of what material the mountain is composed, and measuring its bulk, we are enabled to tell from comparison what is the actual quantity of the material or mass of the earth. Knowing thus the quantity of matter in the earth, and knowing also from astronomical admeasurements its bulk or volume, we can tell its *mean density*. The observation of Maskelyne gave 4.56 for this mean density, and that of Carlini 4.39; which makes the earth's mean density about four times that of water.

ILLUSTRATION II.

EXPERIMENT OF CAVENDISH.—It must be recollected, however, in the above experiment, that the weight of the plummet opposes itself to its *deviation*, and consequently the exact attraction of the mass of matter

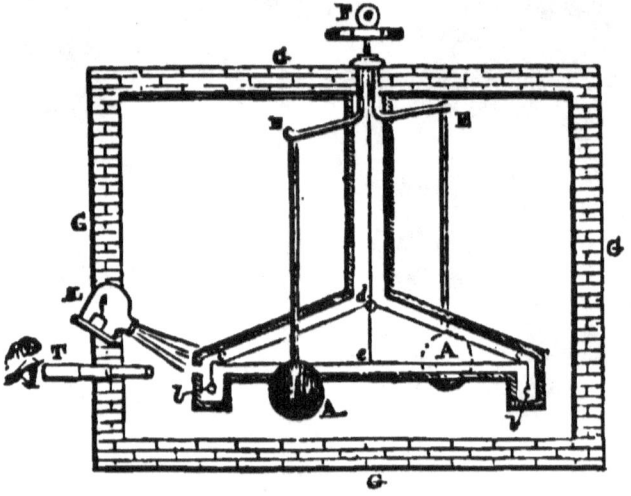

cannot be calculated. But Cavendish, by means of the contrivance to be explained, could easily show the attraction of a less and nearer mass, from which more correct data might be drawn. In this experiment, A A are two balls of lead, weighing about 3 cwt. each, fixed on the extremities of a lever, and capable of being put in motion round an axis above—*b b* are two smaller balls, suspended by slender silver wires from the extremities of a rod *e*, which meet at *d*, and is carried upwards, where, by a delicate arrangement, it is suspended in such a way as to be susceptible of motion by the least conceivable force which it communicates to the rods and its suspended balls. The whole of this last arrangement was suspended in a case marked by the shaded line, that the motion of the balls might be preserved from the impulses of the air, while the whole was enclosed in a room of brickwork G G G G, without door or window, and into which was no other aperture but the one at T, where a telescope lighted by a lamp above, was fixed in the brickwork, by which the extremity of the rod might be seen. The lever or arm which carried the balls A A, were turned by a lever above the chamber at E, and when this arm was thus turned its position was of necessity made to cross that of the light rod cc, carrying the small balls bb; and the greater balls were thus placed in such a position, that their attractions upon the lesser balls should both conspire to turn the rod ee, to which motion of the rod no other force would oppose itself, than the feeble resistance to torsion of the wire ec. The attraction of the *greater* upon the *smaller* balls was sufficient to cause a deflection of the rod de, which, after a *number of oscillations on either side*, at length took up a position nearly in a line, giving the centres of the greater balls, deviating from that position only by the amount due to the torsion of the wire ec. The line of the *oscillation* before the rod eventually rests, being a measure of the attraction of the balls, and sufficient to determine it. Determining, therefore, the attractions of the *greater* balls upon the *less*, he compared it with the attraction of the earth upon these *lesser balls*, and thus he was enabled to compare one mass of the earth with the mass of these greater balls, and knowing the size of the earth, and the size of the balls, he thence obtained a comparison between the *densities* of the two, that is between the densities of the earth, and densities of the lead; he thus found the density of the earth to be 5.48, or about 5½ times that of water. The apparatus of Cavendish is, therefore, *a scale in which the earth, sun, moon, and planets, may be weighed.*

ILLUSTRATION III.

DISSECTION OF CRYSTALS.

Mr. Daniel has thrown much light upon the structure of SOLID BODIES, by the following method of experimenting:—If a lump of alum, borax, or nitre, be immersed in water for three or four weeks, the solution will be found to have gone on unequally. The uppermost portion will be found most wasted, and the undermost least, so that the undissolved parts will have assumed a conical form. The lower part will be found embossed over with numerous crystalline forms. Those in a'um are octahedrons, figures formed by different sections of the aluminous octahedrons. In borax they are fragments of eight-sided prisms, and so on. Hence it follows, that all *these masses are in reality composed of crystals, though such a structure cannot be distinguished by the eye, previous to natural dissection.* The same crystalline structure was developed when carbonate of lime, carbonate of strontian, and carbonate of barytes, were slowly acted on by vinegar; bismuth, antimony, and nickel, treated with very dilute nitric acid, likewise exhibited a crystalline structure, and lime has been crystallized in six-sided prisms, by M. Gay Lussac.

Nature, in various parts of the earth, exhibits crystals of immense magnitude, particularly in basaltic rocks. Basalt is a grey or greenish kind of stone, found in the neighbourhood of present or former volcanoes, and near the sea. These stones have a regular angular shape, and ascend like groups of pillars. Each pillar is formed of many crystals, articulated to each other by joints, that is, each joint is formed of concave and convex surfaces, one being inserted into the other. The Isle of Staffa, or Fingal's Cave, one of the Western Isles of Scotland, is a complete basaltic rock, of a grand and majestic appearance. The drawing will give some notion of this natural wonder, and of the construction of the pillars, some of which are no less than five feet in breadth, and a hundred in height.

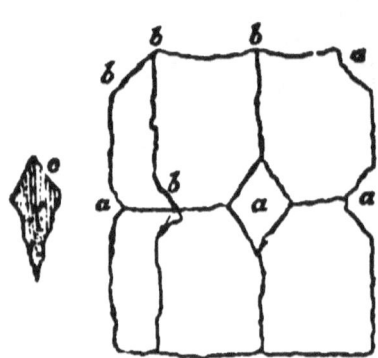

Dr. Brewster, when examining the optical phenomenon of ice, found that even large masses two or three inches thick, formed upon the surface of standing water, are as perfectly crystallized as rock crystal, or calcareous spar; all the axis of the elementary corresponding with the axis of the hexedral prisms, being exactly parallel to each other, and perpendicular to the horizontal surface. This unexpected result was obtained by transmitting polarised light through a plate of ice in a direction perpendicular to its surface. A series of beautiful concentric colored rings, with a dark rectangular cross passed through their centre, were thus exhibited: they were of an opposite nature to those which had been before discovered in the beryl, ruby, and other minerals. The polarizing force of ice was thus found after many experiments to be $\frac{1}{1175}$; that of rock crystal being $\frac{1}{300}$

ILLUSTRATION IV.

BALANCING.

A stick loaded with a weight at the upper extremity can be kept in equilibrio on the point of the finger, with much greater ease than when the weight is near the lower extremity; or, for instance, a sword can be balanced on the finger much better when the hilt is uppermost; because

BALANCING.

when the weight is at a considerable distance from the point of support, its centre of gravity in deviating either on one side or the other from a perpendicular direction, describes a larger circle as at *a*, than when the weight is nearer to the centre of rotation or the point of support as at B. It is according to the same principle that the professed balancer is enabled to perform feats which appear astonishing, as represented in the diagram. It will readily be seen that the difficulty in balancing is not increased with the height of the objects

When a very heavy body is nicely balanced, a very small force is adequate to give it motion. Hence Logan or rocking stones were amongst the *piæ fraudes* of the Druids. Observing so curious a property, they dexterously contrived to

Logan Stone.

make it answer the purpose of an ordeal, and by regarding it as the touchstone of truth, acquitted or condemned the accused by its motions. Mason poetically alludes to this supposed property in the following lines:

" Behold yon huge
And unknown mass of living adamant,
Which, poised by magic, rests its central weight
On yonder pointed rock; firm as it seems,

P

242
MOMENTUM.

Such is its strange and virtuous property,
It moves obsequious to the smallest touch
Of him whose heart is pure—but to a traitor,
Tho' e'en a giant's prowess nerved his arm,
It stands as fixed as Snowdon."

The celebrated Logan Stone here represented, is an immense block of granite weighing nearly sixty tons, and stands near the Land's End in Cornwall. It was a few years since thrown down by some "modern Goths," but the government, much to its credit, ordered it to be replaced.

ILLUSTRATION V.

MOMENTUM.

The following figure illustrates the nature and effects of what is termed momentum. When the figure A is placed upon the step D,

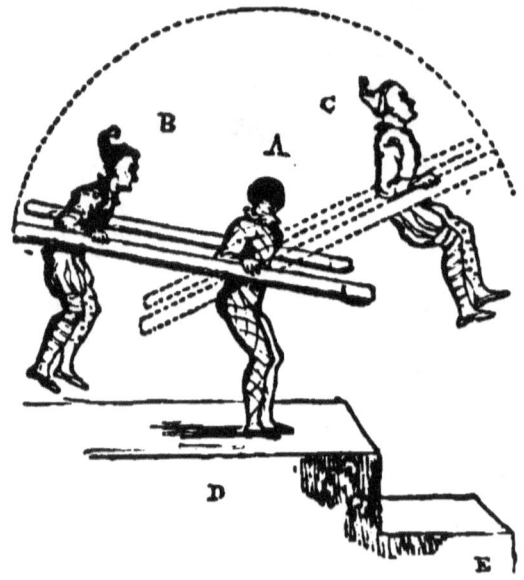

in the position A B, a portion of quicksilver with which the inclined tubes are charged, by running down, swings the figure B round to C; and the centre of gravity having been thus adjusted, the whole would remain at rest, but for the contrivance next described. Besides the connection by means of pivots, the figures are connected with each other by *silken strings*, which keep the figure B steadily in its position, while it traverses the arc until it arrives at C, when their increasing tension has

OBLIQUITY OF MOTION. 243

the effect of capsizing it, and of thus producing a momentum which, by carrying its centre of gravity beyond the line of direction, causes it to descend upon the step E, when the quicksilver, by again flowing to the lowest part of the tubes, places the figures in the same position, only three steps lower, as they were at the commencement of their action, and so on, till any number of steps are descended.

If through a small ball of pith two pins be stuck at right angles to each other, and the points defended with a drop of sealing-wax. It may be then kept in equilibrio at a short distance from the end of a straight tube by means of a current of breath from the mouth. The air blown from the lungs gains such momentum from the contracted channel in which it flows, as to impart considerable velocity to the pins placed within the influence of its current.

ILLUSTRATION VI.

OBLIQUITY OF MOTION.

Every oblique direction of a motion is the diagonal of a parallelogram, whose perpendicular and parallel directions are the two sides. Hence the fall of a battledore in the air, the rifling of a gun-barrel, the feathers of an arrow, and the oblique setting of the vanes of a mill or smoke-jack. When a fire is kindled in the chimney, the air which by its rarification immediately ascends, strikes on the surfaces of the inclined vanes, and by the resolution of forces already explained, causes the spindle to turn round, and gives motion to the spit.

On the same principle an amusing toy may be made. Cut a piece of pasteboard into the following form, and describe on it the spiral line. Cut it out with a penknife, and suspend it on the wire, as seen in the engraving. If the whole be now placed on a warm stone, or under the flame of a candle or lamp, it will revolve with considerable velocity. Some years ago this toy was seen in many shop windows, the *spiral* being sometimes in the form of a serpent, dragon, or other animal.

ILLUSTRATION VII.
CENTRE OF GRAVITY.

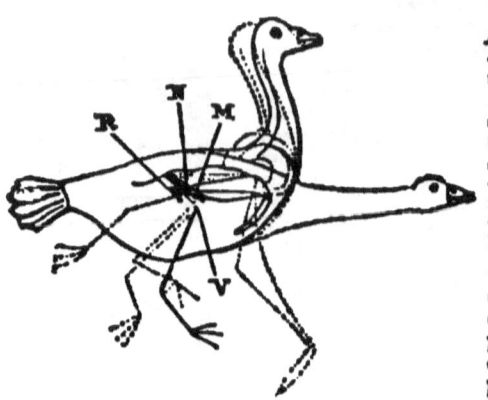

When a body alters its form it changes at the same time the position of its centre of gravity.— In the figure the line drawn from R directs the eye to the position of the centre of gravity, when the bird is *standing*, being then immediately above his foot. When he *swims* the only alteration in his position is the elevation of his legs, accompanied by a corresponding elevation of his centre of gravity, whose position is now shown by the line from N. When he *walks*, his head is thrown a little forward, and his legs are alternately raised, but not so much as in swimming. A more forward and a somewhat lower position must, therefore, be assigned to his centre of gravity,

pointed to by the line from M. When he *flies*, his neck is thrown forward and depressed; his centre of gravity, therefore, advances and sinks, as shown by the line from V.

The stability of a body is increased by lowering its centre of gravity.—This is shown in vibrating figures, as in the following amusing toy. If the ball be removed, the horse would immediately tumble because unsupported, the centre of gravity being in front of the prop; but upon the ball being replaced the centre of gravity immediately changes its position, and is brought under the prop, and the horse is again in equilibrio.

ILLUSTRATION VIII.
GRAVITY.

To cause a body by its own weight to run continually upwards.—The body here is a double cone, which, if placed on two inclined planes that meet at their bases, and afterwards diverge from each other, will immediately put itself in motion and roll up, *because* the centre of gravity of the double cone is in the middle of the line joining its two extremities, and when it is placed between the two inclined planes, the points on which it rests are nearer to the lowest points of the planes than this

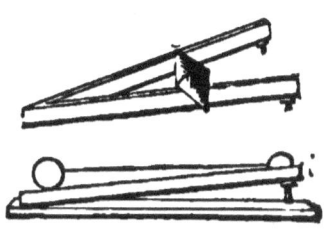

line is. The vertical from the centre of gravity is, therefore, on that side of the points of support which is towards the highest points of the inclined planes. It is in that direction, therefore, that the body has a tendency to roll, and does in reality *descend*, although it appears to *ascend*.

ILLUSTRATION IX.

To cause a cylinder to roll by its own weight a short distance up hill.—The cylinder being loaded at F, near its surface, the position of the *centre of gravity* G may be moved, so that the vertical, G K, from it shall not be from below, but from the point of support at B.

ILLUSTRATION X.

To make a carriage run in an inverted position without falling.—It is a well known fact, that if a tumbler of water be placed within a broad wooden hoop, the whole may be whirled round without the glass falling, owing to the centrifugal force; on the same principle,

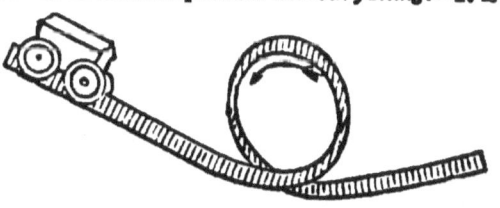

if a small carriage be placed on an iron rail, it will ascend the curve, become inverted, and descend again without falling. A new " London lion," called the *centrifugal railway*, with a carriage sufficiently large to hold two persons is now daily exhibited, by which any one may perform

P 2

the experiment who has no objection to ride for a short time with his head downwards.

ILLUSTRATION XI.

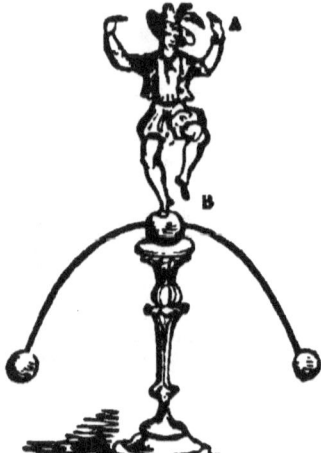

To construct a figure, which being placed upon a curved surface, and inclined in any position, shall, when left to itself, return to its former position.—The feet of the figure A rest on a curved pivot B, which is sustained by the loaded balls below, for the weight of those balls being much greater than that of the figure, their effect is to bring the centre of gravity of the whole beneath the point on which it rests; consequently the equilibrium will resist any slight force given to destroy it.

ILLUSTRATION XII.

ANIMAL MECHANICS.

A bird's wing affords a good illustration of *Animal Mechanics*.

The cut represents the wing of the Ger Falcon. The large bone, marked *a*, is called the humerus, and answers to that part of the human arm between the shoulder and the elbow.

The portion marked *b*, answers to the *fore arm*, or *cubit*, that part of the arm from the elbow to the wrist. This part consists of two bones,—a larger and a curved one, the *radius*, *b*, and a more slender and straighter one, called the *ulna*.

The portion marked *c*, answers to the hand in man: *o* is the *thumb*, which, in the bird, connects what is called the bastard wing. The bones which answers to the palm of the hand, extend from 2, the wrist-joint, to 3, the knuckle-joint.

From the knuckle-joint, at 3, to the point 4, answers to the fingers in the human hand.

The elbow-joint, 1, is a sort of triple hinge, and is so made, that when

MECHANICAL MOVEMENTS. 247

the wing is stretched out, it cannot bend backwards, any more than the pump-handle could fly over the top of a pump. The wrist-joint, also, is on the same construction, and cannot be bent back beyond its proper situation without breaking the wing.

The fore arm having to sustain the broadest part of the wing, is very strong. The *radius*, or larger bone, (which the bird sweeps round in the air,) has the form of a bow; and the smaller one, the *ulna*, the form of a bow-string. It is formed in this manner to afford great strength.

If we take a bow, and having bent it well, place the tips of it on the ground, and then try to bend it downwards, we shall find it much stiffer and stronger than a straight stick of twice the thickness. A very portable bridge is very often made upon this principle. A thin plank is bent into the form of a bow, and its two ends fastened by a cord. When tied in this manner, a plank, which a boy can easily carry, will enable him to cross a chasm ten or twelve feet wide.

But the bow and the plank bridge are strong only in one direction; and, if the position be reversed, they are weaker than a straight stick or plank would be; but the wing of a bird, which has to bear strains and twists in all directions, is formed on a more perfect principle, and will be found, upon examination, to *be a bow in whatever direction it is turned*; and that, in the various motions of the wings, the ulna becomes a tie and a stretcher to the radius in all cases, as they may be required.

ILLUSTRATION XIII.

MECHANICAL MOVEMENTS.

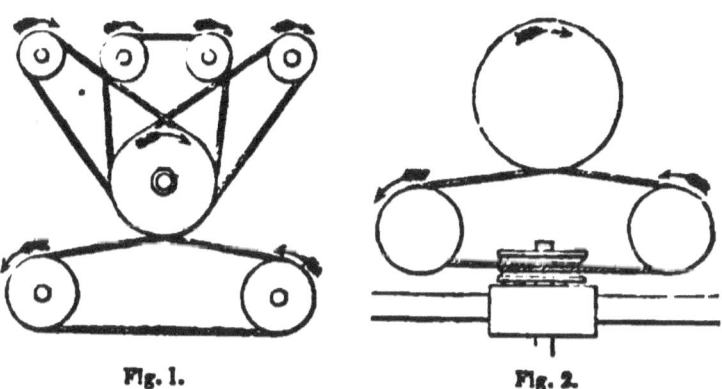

Fig. 1.   Fig. 2.

The above diagrams represent various arrangements for transferring circular motion by means of endless bands, in which the speed is necessarily varied in the proportion of the diameters. In fig 2 horisontal motion is imparted.

248   MECHANICAL MOVEMENTS.

Where circular motion is required to be transferred in a direction not parallel to the driver, bevil wheels are in general used, as in fig. 3.

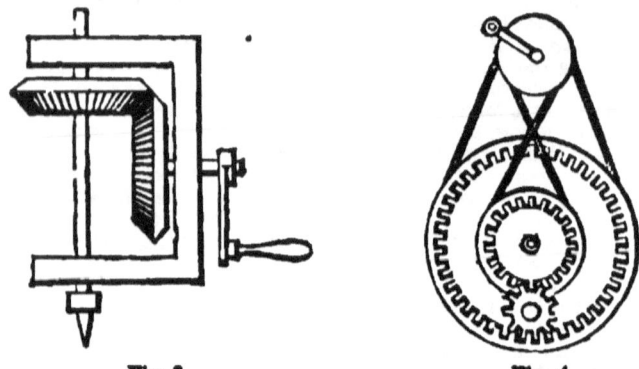

Fig. 3.   Fig. 4.

In fig. 4, suppose the small warve or pulley above to be the driver, the larger wheel with the internal rack, and the concentric wheel within, will be driven in opposite directions by the bands, as represented; and at the same time impart motion to the intermediate pinion, both round its own centre, and also round the common centre of the circular rack and concentric wheel.

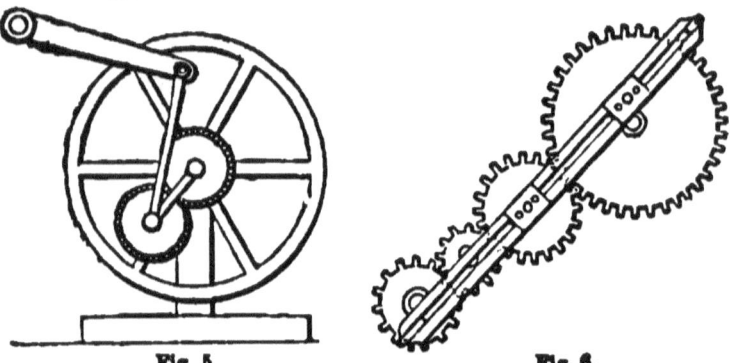

Fig. 5.   Fig. 6.

The arrangement shown in fig. 5, is commonly called the sun and planet wheel, and was first applied, if not invented, by the celebrated James Watt. Two spur wheels are held in gear by a strap or connecting rod from their respective centres, the one being fast in a shaft, and the other fast to the connecting rod, which proceeds from the beam on the left, the vibration of which carries round the fly-wheel, at the same time that the fixed spur wheel in the connecting rod is revolved round the spur wheel 'oto which it gears.

In fig. 6, let the train of wheels be in the proportion of 2, 1, 3, 4, and the sliding part over the second 2, and the 4 fixed at any eccentric points in the respective wheels, the revolution of the wheels will cause the bar to describe curved lines, which may be varied according to the position of the points to which the bar is attached.

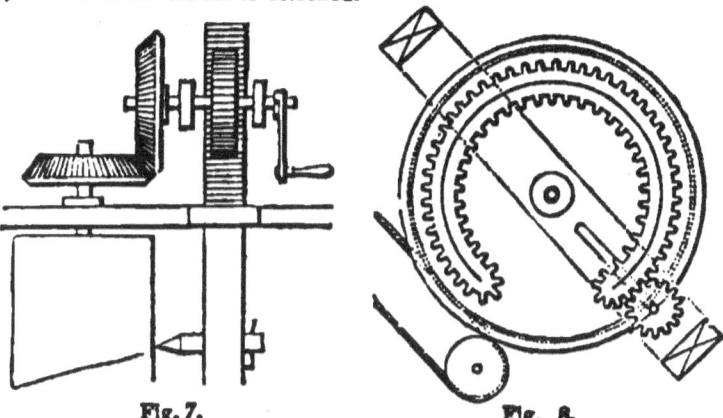

Fig. 7.   Fig. 8.

In fig. 7, suppose the spur wheel which gears into the perpendicular rack to be revolved by the handle on the right, the rack will be moved, and at the same time the bevels will revolve the cylinder with which the horizontal bevel is connected, and a regular spiral line will be described on the surface of the cylinder, by the projecting point connected with the lower part of the rack.

In fig. 8, the pinion having motion on its axis, has also liberty to traverse in the slot, shown by dots, the large double rack will be revolved back and forward on its centre; and according to the ulterior position of the pinion and the band shown at the periphery of the large rack, will produce an alternate rectilinear traverse.

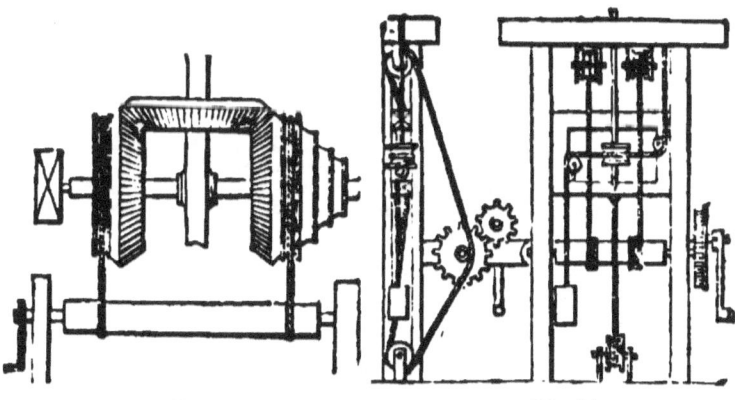

Fig. 9.   Fig. 10.

No. 9 figure on the left hand represents three bevel wheels gearing into each other, that on the left being *fast* on the horizontal shaft, that on the right *loose*, and the upper one loose on the perpendicular shaft, which is held to the horizontal shaft by a circular hole at the common centre of the bevils. Suppose the two bevils on the horizontal shaft to be revolved in *opposite directions* at *equal velocities* by means of the drum beneath, the upper bevel would be revolved in a stationary centre, but supposing the band which revolves the right hand bevel to be removed into one of the smaller pullies with which it is connected, the speed of the two lower bevels will no longer remain the same, and the top bevel will partake of two motions, one round its own centre, another round the centre of the horizontal shaft.

Figure No. 10 exhibits front and side views of a French machine for rifling gun barrels; motion being given to the handle or winch, the sliding carriage which moves perpendicularly in the side guides, is elevated or lowered by the inclined rope in the left-hand figure, at the same time that a rotatory motion is given to the perpendicular boring tool, by another band passing round the horizontal worm, and held tight by means of the weight.

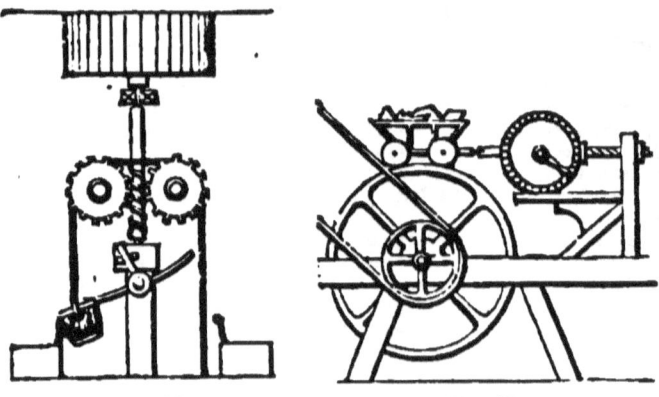

Fig. 11.   Fig. 12.

Suppose the upper part of figure 11 to represent the sails of a horizontal mill, or any sufficient moving power to revolve the shaft which carries the spiral or worm below, and the shaft coupled immediately below the sails, so as to allow of a small vibration, thereby allowing the spiral or worm to act only on one wheel at a time. At the back of these wheels and on the same shafts are placed pullies over which a rope is passed, carrying a bucket at each extremity, one of which is elevated at the same time that the other is lowered by the alternate action of the worm on the opposite wheels. In the centre, and immediately below the worm, is placed a vibrating piece, against which the bucket strikes in its ascent, and which by means of an arm connected with the step in which the the worm shaft is supported, traverses the worm from one wheel to the other, by which means the bucket which has delivered its water is again lowered at the same time that the opposite one is elevated.

The other diagram, fig. 12, represents Mr. Roberts's contrivance to prove that carriages do not present more friction at a great velocity than at a less, if that velocity be once obtained. The upper part of the figure represents a loaded wagon supported on the surface of a wheel, connected to an indicator constructed with a spiral spring, showing the amount of force required to keep the wagon stationary, when the large wheel on which it rests is put in motion. The wagon being loaded, it was found that the number of the revolutions of the large wheel did not vary the effect on the indicator, but that the amount of weight placed in the wagon immediately varied the position of the pointer of the indicator, thus proving that the friction of any carriage on a road does not increase with the speed, but by its weight only.

ILLUSTRATION XIV.

THE RECOIL OF FIRE-ARMS.

The discharge of a shot from a cannon is the result of the elasticity of the gas into which the gunpowder is converted by the application of flame. When this gas suddenly expands, it presses equally in all directions; but the sides of the cannon repel it equally, and therefore neutralize the side pressure. But the breech and the muzzle of the gun are not upon equal terms; the one is a thick mass of iron, the other consists only of the shot and wadding, which accordingly give way. The cannon and the ball receive from the explosion equal forces of motion, the one backwards and the other forwards: the former is the force of the recoil. If the weight of the cannon were only equal to that of the ball, having the same force of motion it would have the same velocity; and the two would fly in opposite directions, equal distances; but as the cannon is heavier than the ball, the same force of motion in it produces greatly less velocity, and the less as the disproportion is greater.

That the recoil of a cannon does not become sensible until the ball has left its mouth, was proved in an experiment made at Rochelle in 1667, by order of the Cardinal de Richelieu. A cannon was fixed in a horizontal position, at the end of a long vertical shaft or rod, moveable freely about an axis at its other extremity. The ball was then fired and struck the object towards which it was directed, precisely as it would have done if the cannon had been fixed, showing that there was no sensible alteration of its position until the ball was discharged from it.

The force of a cannon ball is not communicated to it instantaneously, but by impulses of the air liberated from the gunpowder, which impulses are continually repeated until it finally leaves the barrel. The larger the barrel is, the longer these impulses are continued, and therefore the greater is the accumulated force. In some parts of South America the savages propel small arrows through long slender tubes by blowing into them.

ILLUSTRATION XV.

MILL-WORK.

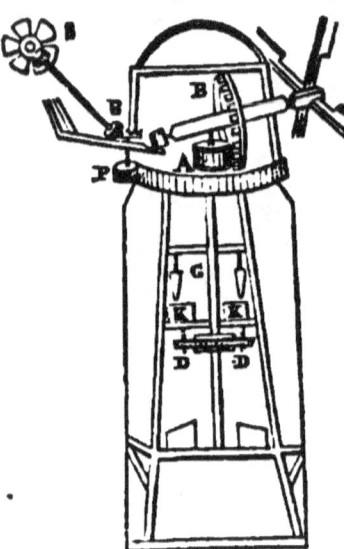

The engraving represents the section of a wind-mill; to the right are the sails, which, by the force of the wind, turn the wheel B, which works upon A; and, by means of the shaft underneath, gives transferred motion to D D, moving the mill-stones K K with corn fed by the bags at G. The machinery, marked by the letters R S P to the left, represent the manner in which the top of the mill is moved round by the wind itself, that its sails may always be presented to it. S is a fan wheel, which by its motion causes the rotation of the cog at R, and is transferred to P, which working into the rack, passing quite round the mill, moves it as required; thus saving the arrangement for hauling the mill round when the wind shifts.

ILLUSTRATION XVI.

FIRE ENGINE.

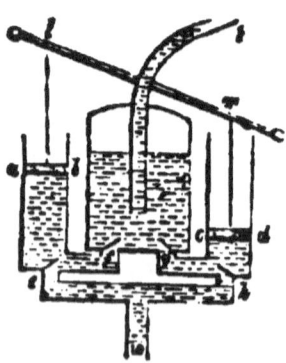

This is of very simple construction, and owes its effect to the action of the forcing pump. *l m* are levers working on a common axis; *a b* and *c d* are two force-pumps; between them is placed a receiver R, called the air-vessel, out of which passes a tube T, the end of which has a metallic spout. The water, therefore, arises from the well or plug *w*, passes into the receiver R, and as it rises in the receiver condenses the air above it, which, by its elastic pressure, forces the water through the tube T.

ILLUSTRATION XVII.

Clock-work.

In page 9 was exhibited the mode of increasing velocity in wheel-work, we now pass to a complete clock, as represented in the engraving. The maintaining power, or weight a, is seen suspended beneath; it acts on the barrel b, to which is attached a large wheel. The teeth of the barrel-wheel give motion to the pinion c, which carries the minute hand d. A small nut at e gives motion to the hour wheel f. Having thus obtained the division of minutes and hours, the next step is to regulate the motion of the train of wheels and pinions. This is effected by the escapement, which consists in the present case of a wheel g, and balance h. The pallets $i\,i$ are intended alternately to take the teeth of the wheel in which they act. This produces the ticking sound so audible in every species of clock. The balance h, with its weights continually oscillates backwards and forwards, and by its slow regular motion, gives uniformity to the operation of the wheel-work of the clock.

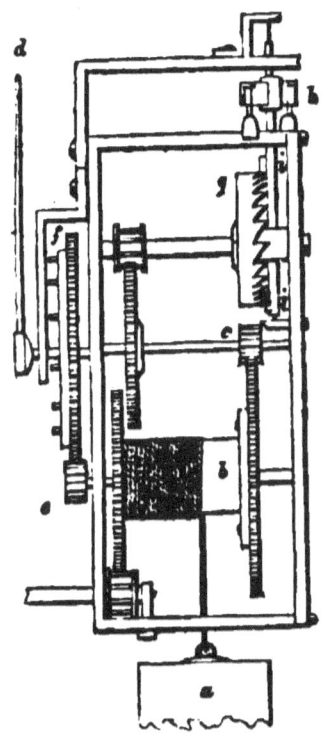

The figure below represents the dead beat of Graham in which the second hand is left in a state of rest between each oscillation of the pendulum.

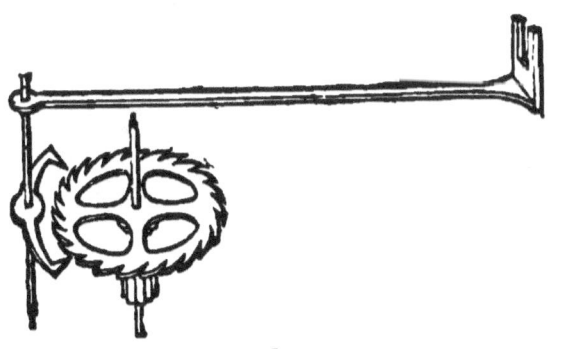

ILLUSTRATION XVIII.

A PORTABLE CLOCK is here represented. In this machine a bent spring is substituted for the weight. The spring is coiled in the barrel k, and the chain being wound round the fuzee gives motion to the train as in the former case. The fuzee is tapered from the bottom upwards. This is intended to equalise the power of the spring, which, when wound quite up, pulls with greater force when its coils are so tightened.

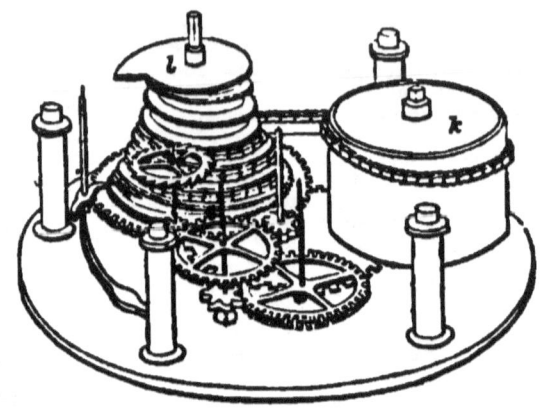

ILLUSTRATION XIX.

The engraving subjoined represents the Printing Machine, and consists of a steel cylinder attached to the reservoir of ink. Upon the solid steel table lie the eight pages of type, forming one side of the sheet, at the top of the machine where the man stands is a heap of wetted paper The machine being set in motion by the boy at the wheel, the boy behind places a sheet upon a flat table before him, which is seized by the great drum or cylinder, at the same moment the inking-apparatus is set in motion. The sheet is now caught by the wet roller, and conveyed over the first cylinder, and immediately receives an impression from the type, and this process is repeated with inconceivable rapidity.

PNEUMATICS AND HYDRAULICS.

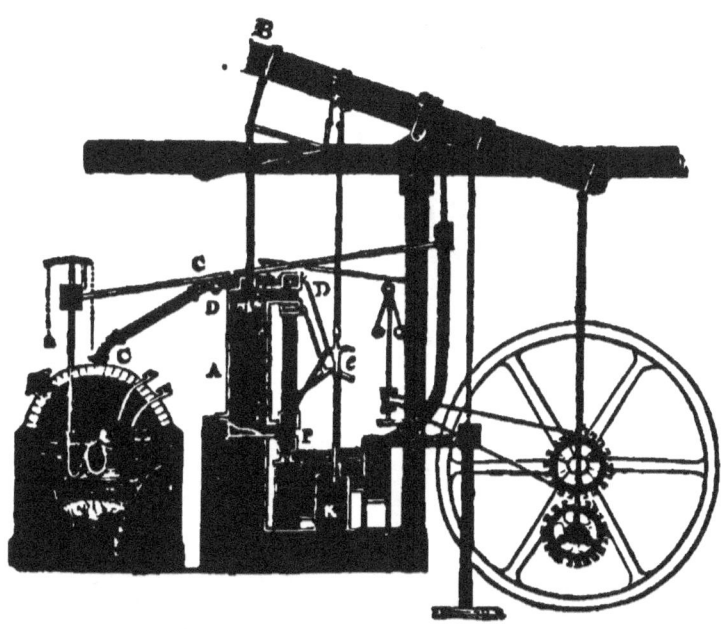

ILLUSTRATION I.

STEAM ENGINE.

The Steam Engine is one of the most wonderful applications of the power of machinery in the present time. The first idea of it is ascribed to the Marquis of Worcester; there is, however, good reason for supposing that Brança, an Italian, proposed turning the vanes of a mill by a steam kettle, and to use steam as an impelling power for a stamping engine, so early as 1629. The hint of the Marquis of Worcester was given in the Century of Inventions in 1663, and was carried into effect by a person named Savory, a captain of a merchant vessel, in 1691. It has since that period been greatly improved by various persons by Newcomen and Crawley, but more especially by Watt.

ILLUSTRATION II.

Steam itself appears to have been employed as a prime mover by Hiero, a Greek, more than 2,000 years ago; and the early philosophers had many modes of producing a rotatory motion with highly elastic vapour, as in the following figure which represents a hollow globe, having four tubes opening into the atmosphere, opposite each other, through which the steam made its escape, and gave motion to the machine.

ILLUSTRATION III.

Brança formed a boiler, which he represented in the shape of a human head, to which he applied heat. The steam or vapour rushed out of the figure, and gave motion to a horizontal wheel, as seen in the engraving.

ILLUSTRATION IV

SAVORY'S ENGINE.

This apparatus consisted of a boiler O, furnished with a safety valve acting by means of a weight W, at the end of the lever. The steam vessel S, into which the steam was made to pass through the pipe N,

ATMOSPHERIC ENGINE. 257

was connected with the well C, by the suction-pipe H P, and when water was to be raised the vessel S was filled with steam, which completely expelled the air. The communication with the boiler was then closed by shutting the valve Q, and the steam in the vessel becoming condensed, a vacuum was formed, into which the water rushed from the well by the pressure of the atmosphere on its surface. The handle R is used to give motion to the double valve I and Q, the former being intended to admit condensing water from the pipe A D, while the latter alternately opens and shuts the steam communication. In this form of the apparatus the inventor was seldom able to raise water more than thirty feet, and when a greater altitude was required, it was effected by the impelling force of the steam. This was accomplished by the ascending pipe A D, which was sometimes carried sixty feet higher than the steam vessel S. After condensing the steam, and filling the vessel S with water, a new supply of steam was introduced, which, pressing on the surface of the water, drove it up the pipe D; but, as we have just stated, it will be evident that this operation must be sometimes very dangerous, as the pressure on the internal surface of the boiler must be in proportion to the height of the column of water raised by the elasticity of the steam.

ILLUSTRATION V.

ATMOSPHERIC ENGINE.

In 1707 the steam engine was greatly improved by Newcomen and Crawley. They constructed what is called the atmospheric engine. In this the steam is generated in the boiler B, and is admitted into the cylinder A, when the piston is forced to the top as represented in the engraving. D D is a large beam or lever, moving on an axis, and supporting at one extremity the rod E of the steam piston, by means of a chain passing over an arch of wood, at the end of the lever or working beam. The reason for employing this arch will be obvious, if we consider that the distance of the extremity of a common lever from a perpendicular beam supporting the fulcrum is continually varying, and would therefore draw the piston rod out of its vertical direction; but by attaching a flexible chain to part of a circle of wood, of which the fulcrum is the centre, the piston rod is constantly kept at the same distance from the perpendicular beam, and its vertical motion is preserved. The air being expelled by the admission of the steam to the cylinder A, (which is placed within an exterior cylinder), the communication with the boiler is closed, and a portion of cold water, admitted outside the cylinder from the reservoir H, instantly condenses the steam, and forms a vacuum below the piston, which is then forced to the bottom of the cylinder by the pressure of the atmosphere.

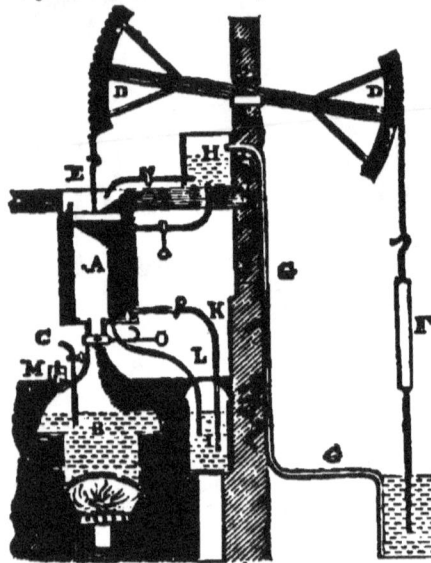

To the opposite end of the lever beam the pump rod is suspended, furnished with a counterpoise F, to elevate the piston rod after its descent in the cylinder. The piston being thus alternately elevated and depressed, a quantity of water will be raised from the well at every stroke of the engine, in proportion to the force with which the atmosphere presses upon the piston. M is a safety valve, and C a small pipe descending into the water in the boiler, for the purpose of ascertaining the quantity it contains: for if the cock at C is opened, while the lower extremity of the pipe is immersed in the water, the elasticity of the steam will force water through the aperture; but if the surface of the water in the boiler is below the end of the pipe, only steam will issue when the cock is turned. G G is a tube communicating with the well to supply condensing water to the reservoir. H, K and L are pipes for conveying the water, formed by the condensation of the steam on the interior and exterior of the cylinder, into the well I.

ILLUSTRATION VI.

DOUBLE-ACTING ENGINE.

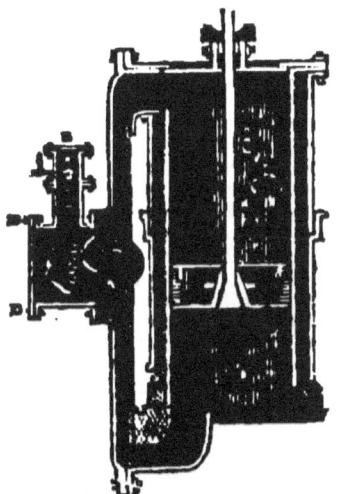

Watt was originally a mathematical instrument maker in the city of Glasgow. In 1765 he was employed to repair the model of Newcomen's steam engine, belonging to the university of that place; and while examining this imperfect apparatus, he observed the immense loss of steam occasioned by its admission to the cylinder, just cooled for condensation. He found, that a certain number of cubic inches of steam, generated in the boiler, did not operate as they ought in producing a vacuum. In fact, he ascertained by experiments, that one half of the steam of the boiler was lost by its being condensed in the cylinder previously cooled.

He therefore resorted to several contrivances in order to prevent this condensation. He first made a wooden bottom to the piston, and then made the cylinder wholly of wood. He afterwards enclosed it with a packing of hemp, to prevent the escape of heat by radiation, and finally introduced the important improvement of condensing the steam in a separate vessel, by which means the cylinder was constantly preserved at the boiling temperature. His earliest model is constructed upon this admirable principle, in the practical development of which, it is stated, he received considerable assistance from the celebrated Dr. Black, whose discoveries on the subject of heat then excited the general attention of the scientific world. The first improvement then made by Watt, consisted of the addition of a condensing cylinder, into which the steam, after raising the piston, was allowed to pass, by opening a valve, where it condensed without lowering the temperature of the working cylinder. The next great improvement he made was that of making the steam itself depress the piston instead of the pressure of the atmosphere. This was effected by admitting the steam both above and below the piston, in a manner which may be understood by an examination of the diagram.

When the piston is merely pressed down by the weight of the atmosphere, it is obvious that the pressure cannot be greater than about 15lbs. to the square inch; but by admitting steam above the piston, and thus effecting its depression independent of atmospheric pressure, as in the above diagram, the force with which it is driven down can be increased to an almost unlimited extent.

The cylinder A is furnished with a steam-tight piston, the rod of which is connected with the working beam. This cylinder, instead of being open at the top, like that in Newcomen's steam engine, is air-tight. The steam is admitted from the boiler through a pipe, of which the orifice is seen at B. Below this opening is situated the throttle valve c, which regulates the quantity of steam admitted through the steam pipe, by closing or opening in proportion to the divergency of the governor, with which it is connected, in a manner described in the diagram, "governor balls." The steam passes through the box D D, and enters the cylinder either above or below the piston, according to the situation of the sliding valve, or bridge a, contained in the box, which being elevated or depressed by the motion of the toothed arch, working into the teeth of the valve, alternately opens and closes the apertures which communicate with the upper and lower parts of the cylinder. In the figure the lower aperture F is open to the steam box, and the steam is entering beneath the piston, in the direction shown by the arrow, and at the same time a communication is formed from the upper part of the cylinder through the appertnre m, to the pipe through which the steam passes into the condenser. When the piston has reached the top of the cylinder, the direction of the sliding valve is changed, so that the pipe m opens into the steam box, and the elastic vapour is admitted above the piston, while the aperture F is placed in communication with the condenser, into which the steam passes from the lower part of the cylinder during the descent of the piston. By this admirable contrivance, the upper and lower parts of the cylinder were alternately made to communicate with the steam box and the condenser, and the piston being both elevated and depressed by the power of steam alone, the machine in the hands of Mr. Watt became a real steam engine. In the figure, (illustration 1,) the steam passes through the pipe C C to the box D; the lower aperture of the cylinder is open at P, and the steam is entering beneath the piston, at the same time a communication is formed from the upper part of the cylinder through an aperture to the pipe through which the steam passes into the condenser K.

ILLUSTRATION VII.

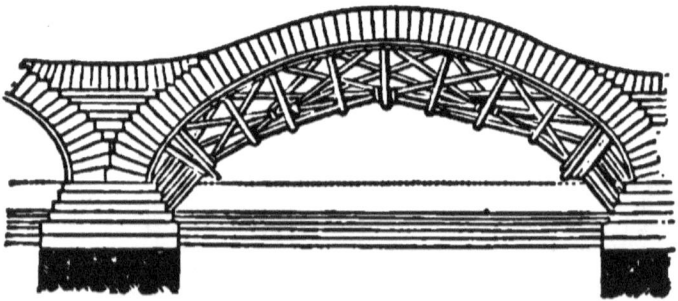

The Arch—Method of centering.

RAILWAY ENGINE. 261

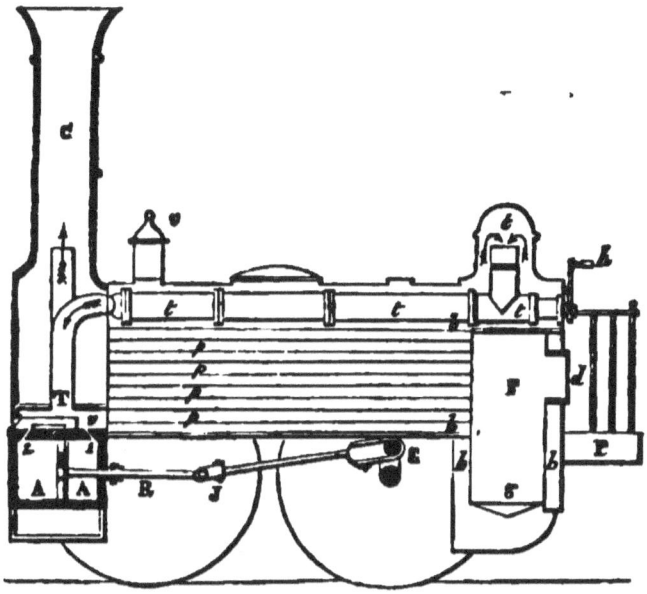

The plate represents the longitudinal section of an engine, and is designed to give a general notion of the manner in which it is impelled, without entering into any thing like the descriptive minutiæ of the machinery. In the first place then, F is a square furnace, surrounded on all sides, except d the door, and g the grate, with b, b, b, b, the boiler; from the furnace F, through the boiler, run a number of pipes or flues p, p, p, p, communicating with the chimney C. At the top of the boiler is a large bent tube t, t, t, which opens into the boiler at t, where the tube is so elevated that the motion of the engine cannot splash the water up into it. At t, is a regulating valve or cock, which by the handle h, can open a communication from the top of the boiler down the tube, or stop it at the will of the engineer, who stands on the platform P, just by the handle. At T, the tube terminates in the valve box v, out of which are two openings 1, 2, into the cylinder A, A, on the top of these run a sliding valve S, which is so constituted as to be able to cover only one of the entrances to the cylinder at a time; it is slided backwards and forwards by the machinery when in action, and thus covers and uncovers the entrances 1 and 2 alternately. B is the piston which moves steam tight to the left and right in the cylinder; from it the piston rod R runs out through an aperture in the side of the cylinder, in which aperture it plies backwards and forwards steam tight; but to enable it to turn the crank E, it is necessary that it should bend upwards and downwards; to effect this, a joint is made at J. The fire being lit, and the boiler filled with water, nearly up to the large tube t, the engineer from time to time throws in coke through the

Q 3

door *d*, and the combustion increasing, an intense heat is kindled, and acting on the sides of the furnace and passing by the small flue pipes through the boiler *b*, is communicated to the water through the sides of the furnace and pipes. The water, heated from so many surfaces, becomes powerfully excited, boils and changes into volumes of steam, which soon fill the upper portion of the boiler, where, being limited in space, it becomes compressed, and thus gains an immense expansive force, and acts on all sides with a permanent giant effort at the disruption of its prison; at this crisis the engineer turns the handle A, and the steam instantly explodes down the tube *t, t, t,* in the direction of the darts, and rushing through the valve box, passes the aperture 1, enters the right side of the cylinder A, and pressing on all sides of it, forces the piston to the left, by which action the piston rod is drawn in, and the wheel W turned by the agency of the crank E, and the engine moves on the rails; but the motion of the piston from right to left, by the agency of machinery, is made to move the slide in the valve box, from left to right, and thus the aperture 1 is closed, the aperture 2 is opened, and the pressure of the steam transferred from the right side of the piston to the left, when it is propelled back to the right, and this motion sending back the sliding valve to the left, the steam again rushes in through the aperture 1: thus the expansive force of the steam alternately presses on each side of the piston, and thus the motion of the engine is continued. When the piston is forced from one side to the other, the part of the cylinder that it has left is of course filled with steam, this must be let out to permit the piston to come back again, and having performed its function, it is emitted into the chimney by a valve worked by the machinery, and passes off with the smoke.

OPTICS AND ASTRONOMY.

ILLUSTRATION I.

THE THAUMATROPE.

This word is derived from two Greek words $\theta\alpha\upsilon\mu\alpha$ and $\tau\rho\epsilon\pi\omega$, the former of which signifies *wonder*, and the latter to *turn*. This philosophical toy is founded upon the well-known optical principle, that an impression made upon the retina of the eye, lasts for a short interval after the object, which produced it, has been withdrawn. The impression which the mind thus receives lasts for about the eighth part of a second, as may be easily shown by whirling round a lighted stick, which, if made to complete the circle within that period will exhibit not a fiery point but a fiery circle in the air.

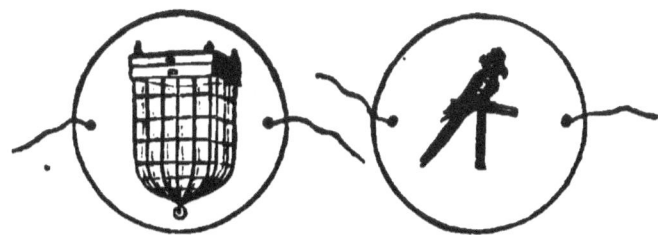

If a circular card be painted with a bird on one side and a cage on the other, and two strings be fastened on its axis, by which the card can be easily made to revolve by means of the finger and thumb, the bird will appear to be caged. A bat may in the same manner be painted on one side of the card, and a head upon the other, which will exhibit the same phenomenon.

On the same principle as that upon which the thaumatrope is founded, another ingenious toy is constructed. It consists of a disc of darkened or japanned tin-plate, with a slit or narrow opening as seen in the cut. If a device of any kind, such as a star, be painted on a card and fixed upon a pin, and made to revolve behind the disc, and the eye be placed at the opening the whole will be visible. If, instead of the star a device, such

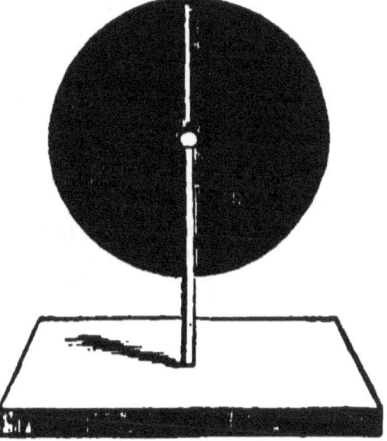

as that in the cut, viz., a number of boys, in the attitude of leaping be introduced, and the instrument arranged below a looking-glass, the whole of the figures will appear in motion as if leaping up and down, increasing in agility as the card revolves with more rapidity.

ILLUSTRATION II.

TELESCOPES.

The invention of the telescope is ascribed to different persons, among whom are John Baptista, Ports Jausen of Middleburg, and Galileo. The time of its construction is considered to have been about 1590.

The structure and arrangement of the glasses in a telescope will be best understood by a reference to a series of figures.

Let c d be a convex glass fixed in a long tube, and having its focus at E; then a pencil of rays, g h i flowing from the upper extremity, A, of the remote object A B, will be so refracted by passing through the glass as to converge and meet in a point f, whilst the pencil of rays k l m, flowing from the lower extremity B, of the same object A B, and passing through the glass, will converge and meet in the point e; and the images of the points A and B will be formed in the points f and e; and, as all the intermediate points of the object between A and B send out pencils of rays in the same manner, a sufficient number of these pencils will pass through the object glass, c d, and converge to as many intermediate points between e and f, and so will form the whole inverted image e E f, of the distinct object. But because this image is small, a concave glass n o, is so placed at the end of the tube next the eye that its virtual focus may be at F; and as the rays of the pencils pass converging through the concave glass, but converge less after passing through it than before, they go on further, as to *b* and *a*, before they meet, and the pencils themselves being made to diverge by passing through the concave glass, they enter the eye and form the large picture *a b* upon the retina, whereon it is magnified under the angle *b* F *a*.

But this telescope has one inconvenience which renders it unfit for most purposes, which is that the pencil of rays being made to diverge by passing through the concave glass *n o*, very few of them can enter the pupil of the eye, and therefore the field of view is but very small, as is evident by the figure; for none of the pencils which flow either from the top or bottom of the object A B can enter the pupil of the eye at C, but are stopped by falling upon the iris above and below the pupil, and therefore only the middle part of the object can be seen when the telescope lies directly towards it, by means of those rays which proceed from the middle of the object; so that, to see the whole of it, the telescope must be moved upwards and downwards, unless the object be very remote, and then it is never seen distinctly.

This inconvenience is remedied by substituting a convex eye glass, as *g h* in the annexed figure, in place of the concave one, and fixing it so in the tube that its focus may be coincident with the focus of the object-glass *c d*, as at E; for then the rays of the pencils flowing from the object A B, and passing through the object glass *c d* will meet in its focus and form the inverted image *m* E *n*; and, as the image is formed in the focus of the eye glass *g h*, the rays of each pencil will be parallel, after passing through that glass; but the pencils themselves will cross in its focus on the other side, and, the pupil of the eye being in this focus, the image will be viewed through the glass under the angle *g e h*, and being at E it will appear magnified so as to fill the whole space, C *m* E *n* D.

A telescope for viewing terrestrial objects should be constructed so as to show them in their natural posture; and this is done by employing one object glass *c d*, represented in the subjoined figure, and three eye glasses, *e f*, *g h*, *i k*, so placed that the distance between any two, which are nearest to each other, may be equal to their focal distances, as in the figure, where the foci of the *c d* and *e f* meet at F; those of the glasses *e f* and *g h* meet at *l*; and of *g h* and *i k* at *m*, the eye being at *n*, in or near the focus of the eye glass *i k*, on the other side. Then it is plain that those pencils of rays which flow from the object A B, and pass through the object glass *e d* will meet and form an inverted image C F D, in the focus of that glass; and, the image being also in the focus of the glass *e f*, the rays

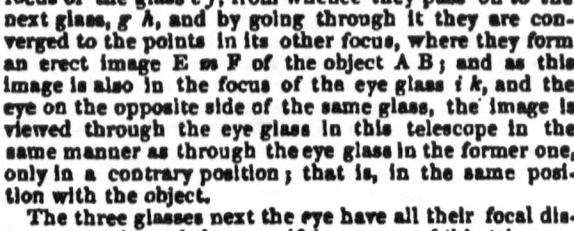

of the pencils will become parallel, after passing through that glass, and cross at *l* in the focus of the glass *e f*, from whence they pass on to the next glass, *g h*, and by going through it they are converged to the points in its other focus, where they form an erect image E *m* F of the object A B; and as this image is also in the focus of the eye glass *i k*, and the eye on the opposite side of the same glass, the image is viewed through the eye glass in this telescope in the same manner as through the eye glass in the former one, only in a contrary position; that is, in the same position with the object.

The three glasses next the eye have all their focal distances equal; and the magnifying power of this telescope is found in the same way as that of the last described, viz., by dividing the focal distance of the object glass *c d*, by the focal distance of the eye glass *i k*, or *g h*, or *e f*, since all these three are equal.

When the rays of light are separated by refraction they become coloured, and if they be united together again they will be a perfect white; but those rays which pass through a convex glass near its edges are more unequally refracted than those nearer the centre of the glass; and, when the rays of any pencil are unequally refracted by the glass, they do not all meet again in one and the same point, but in separate points, which make the object indistinct and coloured about the edges. The remedy is to have a plate with a small round hole in its middle fixed in the tube at *m*, parallel to the glasses; for the wandering rays about the middle of the glasses will be stopped by the plate from coming to the eye, and none admitted but those which come through the middle of the glass, or at least at a good distance from its edges, and pass through a hole in the middle of the plate. But this circumscribes the image, and lessens the field of view, which would be much larger if the plate could be dispensed with.

The most important improvement in this instrument consists in the formation of the object glasses, free from the errors of chromatic and spherical aberration, whence they have been denominated *achromatic* telescopes, or more properly *aplanatic* telescopes. These are now made in such perfection, that they have, in some degree, superseded the reflecting telescopes. Dolland first made achromatic telescopes; they are formed by employing a compound object-glass, composed of a series of lenses of different refractive powers, which will mutually correct each other, and thus give a pencil of light entirely colourless.

ILLUSTRATION III.

REFLECTING TELESCOPE.

The reflecting telescope was invented by Sir Isaac Newton, and improved by Dr. Gregory, and is generally preferred for astronomical purposes, because the principle of its construction admits of its being made to magnify more than a refracting telescope of the same length. It has also the advantage of being perfectly *achromatic*, that is, it produces no coloured or rainbow edges to the images.

The following figure, (Fig.21,) will lead to a description of one of these most in use. As there is a great similarity between *convex lenses* and *concave mirrors*, both forming an inverted focal image of any remote object, by the convergence of the pencil of rays, in these instruments the concave mirror is substituted for the convex lens. T T represents the large tube, and *t t* the small tube of the telescope, at one end of which is D F, a concave mirror, with a hole in the middle at P, the principal focus of which is at I K; opposite to the hole P is a small mirror L, concave towards the great one; it is fixed on a strong wire M, and may, by means of a long screw on the outside of the tube, be made to

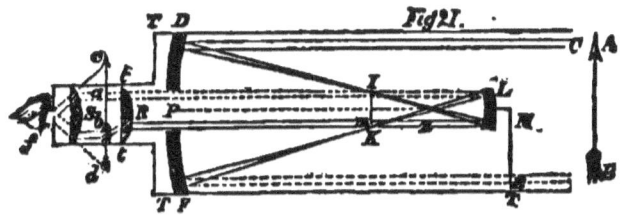

move backwards or forwards. A B is a remote object, from which rays will flow to the great mirror D F. There are only two rays of a pencil taken from the top, and two from the bottom; and, in order to trace the progress of the reflections and refractions, the upper ones are represented by full lines, the lower ones by dotted lines. Now the rays at C and E falling upon the mirror at D and F, are reflected, and form an inverted image at K I. As there is nothing there to receive the image, the rays go on towards the reflector L, the rays from different parts of the object crossing one another a little before they reach L. From the mirror L the rays are reflected nearly parallel through P; there they have to pass the plano-convex lens R, which causes them to diverge at *a b*, and the image is now painted in the small tube near the eye. Having by means of the lens R, and the two concave mirrors brought the image of the object so nigh as at *a b*, we only want to magnify the image, and this is done by the lens S; and it will appear as large as *c d*, that is, the image is seen under the angle *c f d*.

ILLUSTRATION IV.

MICROSCOPE.

The MICROSCOPE is an instrument for viewing very small objects; the principle of its construction, is the bringing of the object to increase the angle under which it is seen, and thereby apparently to enlarge it. If it be an object not clearly visible at a less distance than A B, and it be placed

in the focus C of the lens D, the rays which proceed from it will become parallel by passing through the said lens, and therefore the object is distinctly visible to the eye at E. There are four descriptions of microscopes, the *single*, the *compound*, the *solar*, and the *oxyhydrogen* microscope, for the latter, see "Illustrations of Chemistry."

A very convenient form of microscope is shown in the following engraving, where A is a circular piece of brass or ivory, in the middle of which is a small hole, one-twentieth of an inch diameter: in this hole is fixed, with a wire, a small lens, whose focal distance is *c*. At that point is placed a pair of pliers, D, which may be adjusted by means of the sliding screw, as in the figure, and opened by means of two little studs. The object may be viewed 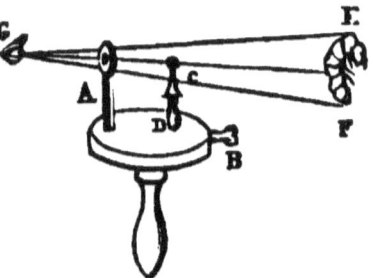 with the eye placed in the other focus of the lens at A; and, according to the focal length of the lens, the object C will appear more or less magnified, as represented at E F. If the focal length be half or one-fourth of an inch, the length, surface, and bulk of the object will be magnified in a similar proportion. This small instrument may be put into a case for the pocket. Those lenses whose focal lengths are three-tenths, four-tenths, and five-tenths of an inch, are the best for common use.

SOLAR MICROSCOPE.

The following diagram exhibits a double or compound microscope. *c d* is called the object glass, and *e f* the eye glass. The small object *a b* is placed a little farther from the glass *c d* than its principal focus, so that the pencils of rays, flowing from the different points of the object, and passing through the glass, may be made to converge and unite in as many points between *g* and *h*, where the image of the object will be formed. This image is viewed by the eye glass *e f*, which is so placed, that the image *g h* may be in the focus, and the eye at about an equal distance on the other side; the rays of each pencil will be parallel, after going out of the eye glass, as at *e* and *f*, till they meet at K.

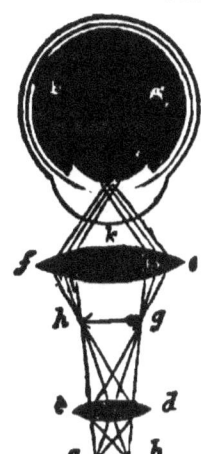

SOLAR MICROSCOPE.

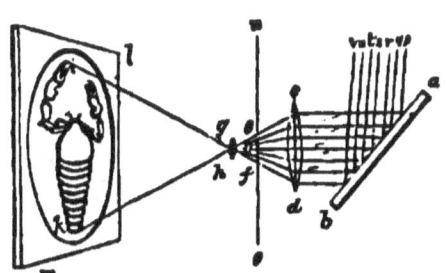

This instrument is shown in its simplest form in the above engraving, in which *a b* is the diagonal mirror for receiving the rays of light, *p q r s t u v*. They are reflected by the polished surface, and thrown on the lens *c d*. Within the focus at *e f* is placed any transparent object to be magnified, and the image thus illuminated passes through the lens *g h*. The size of the magnified figure, *i k*, will depend on the distance the instrument is placed from the wall *l m*.

ILLUSTRATION V.

MURAL CIRCLE.

This engraving represents one of those magnificent instruments called a mural circle, with the position of the observer.

The circles at Greenwich are each of them six feet in diameter, and are connected with their centres by sixteen conical radii, in the same manner as the spokes of a wheel connect the rim with the nave; indeed, the whole instrument very much resembles, both in shape and motion, a coach wheel. The telescope is attached to the face of the circle, and in its inside, near the end to which the observer places his eye, are fixed two spiders' threads, drawn very tight across the diameter, and at right angles to each other; their intersection takes place in (and denotes) the centre of the telescope; these spiders' lines are so magnified by the eye-piece of the instrument, that, fine as they are, they appear like stout black threads; anything really stouter would be much too thick when viewed in the telescope.

The axis on which the circle turns passes through a hole in the massive stone pier which supports the instrument, and is fastened on the other side. The divisions of the circle into degrees, minutes, &c., are placed round the edge of the rim, and are so exquisitely fine, that they require a microscope of a peculiar construction to read them off; there are six of these microscopes, placed at equal distances round the circle, and are lettered A, B, C, &c. The instrument is so fixed as to turn round in one position only, namely, in the plane of the meridian; that is to say, the face of the circle always stands north and south, and the observer waits for the heavenly bodies coming to the meridian (or what is commonly known as their southing) before he makes his observation; he can then measure both their meridional distance from each other, and their distance from the celestial pole, and hence he deduces their declination.

ILLUSTRATION VI.

TRANSIT INSTRUMENT.

Our next engraving represents the transit instrument, which is ten feet long, and is also fixed so as to move only in the plane of the meridian. A clock, of a superior construction, is placed in view of the observer, to note the instant a celestial object passes the centre of the telescope, (which centre is denoted by spiders' threads), and, if the object be a star whose right ascension is known, the error of the clock is at once found; for, if the clock were quite right, it would indicate at that instant the right ascension of the star; therefore, what it differs from it will be the error of the clock. It will now be understood, that one capital use of this instrument is to determine exact time; another is, to determine the right ascension of all the heavenly bodies, which, together with their declinations found by the mural circles, fixes their relative situations.

CHEMISTRY.

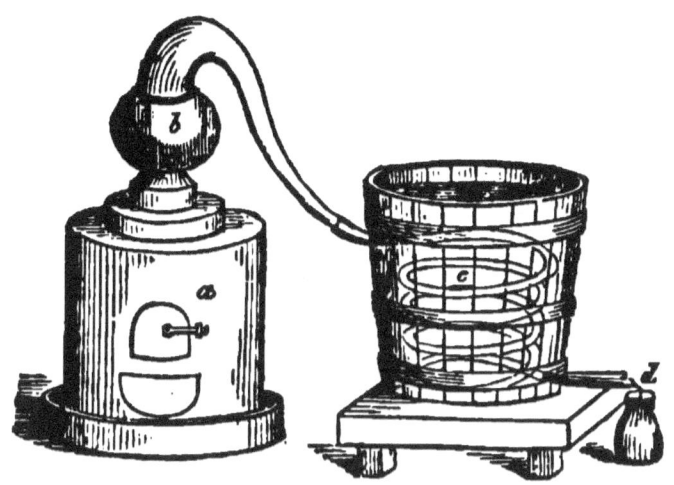

ILLUSTRATION I.
DISTILLATION.

Distillation is that process by which the volatile particles of a body or of fluids are vaporised, condensed, and collected in appropriate vessels. It is performed by submitting liquids to heat, in close metallic or glass vessels. The more volatile particles being in a state of vapour ascend in the body of the still, or lower vessel, in which the liquid boils, and enter the head or capital b; from thence the vapour passes to a metallic tube, called the worm or condensing tube, which is cooled externally by means of water contained in a large tub or wooden vessel called the refrigerator, (c.) The vapour is thus condensed or reduced to a liquid state, by parting with its heat through the metal to the surrounding cold water, and descends through the worm into any convenient vessel below, called the receiver, (d.) a is the furnace on which the still is placed.

ILLUSTRATION II.
AFFINITY OF FLUORIC ACID FOR SILICA, EXEMPLIFIED IN THE ART OF ETCHING UPON GLASS.

Procure some thick clean pieces of crown glass, and immerse them in melted wax, so that each may receive a complete coating; when perfectly cold, draw on them with a fine steel point, flowers, trees, houses, portraits, &c. Then immerse the pieces of glass one by one in a square leaden box or receiver, where they are to be submitted to the action of fluoric acid, or fluoric acid gas.

It will be necessary to have some water in the receiver for the absorption of the superabundant gas, and the receiver should have a short leaden pipe attached to it, for the reception of the beak of the retort; this should be well luted with wax. At the top of the receiver there is a sliding door for the admission of the plates; this is to be well luted while the gas is acting. When the glasses are sufficiently corroded they are to be taken out, and the wax is to be removed by wetting them in warm water. Various colours may now be applied to the glass, whereby a fine painting may be executed. In the same manner initials and sentences may be etched on wine glasses, tumblers, &c. In this experiment the *potash* of the glass is set free, while the silex is acted on. The following is the description of the figure illustrating this experiment, at the head of the chapter on Chemistry, page 183. A, lamp stand; B, argand lamp; C, leaden retort; D, a leaden lengthened tube inserted in the bottom of the receiver; E, the leaden receiver above which the unwaxed parts of the glass are corroded; F, a stand; G G, two pieces of glass; H, a double bottom, the upper one being pierced with holes for the transference of the gas.

Papin's digester is represented by the following cut. It is constructed on the principle of mechanical pressure being necessary to elevate fluids to a higher temperature than those common boiling points. A is the boiler, B the lid, C a valve to allow the escape of a small portion of steam to prevent the bursting of the apparatus, D is a notched lever on which a weight E hangs to prevent the valve from rising by a slight expansion of the steam. In this machine animal bones are dissolved with great facility, in order that the gelatine contained in them may be converted into rich soups.

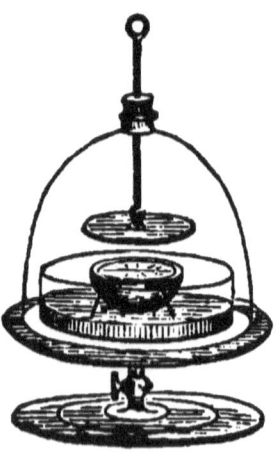

ILLUSTRATION III.

LESLIE'S PROCESS FOR ARTIFICIAL CONGELATION.

The water in a porous pan is set above a wide bason, containing strong sulphuric acid or parched oatmeal. The apparatus being now screwed to the plate of an air pump is exhausted. Whilst the cup is covered the water continues fluid, but on drawing up the lid it freezes instantly. When the action is intense the congelation sweeps at once over the surface of the water like a cloud. At other times the ice is formed in small feathered spiculæ.

ILLUSTRATION IV.
OXYHYDROGEN MICROSCOPE.

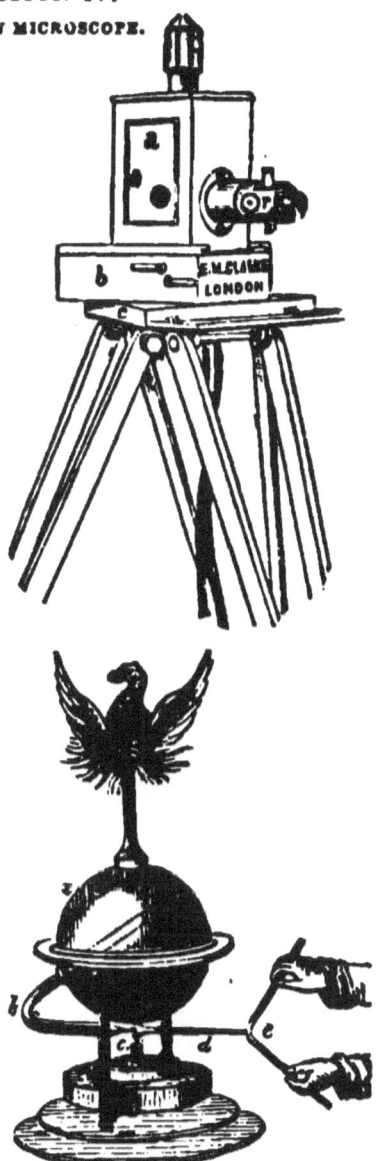

The application of mixed gases for the production of an intense light, is one of the great discoveries of modern times. In the oxyhydrogen microscope, the oxygen and hydrogen gases are retained in separate gas holders, not shown in the figure, which are constructed so as to regulate the proper discharge of each. The two gases are let in by stop cocks at the side of the pillar, and are mixed in a little reservoir at the top of it. After the gases are mixed, they are passed through three screens of wire gauze, to prevent accidents from explosion, and then they give their united action upon a small sphere of pure lime, kept in a state of rotation, when the most brilliant light is evolved. This passes through the microscopic objects under examination, and their images are thrown upon a disc 60 feet in diameter, being often magnified upwards of 50,000 times.

The same principle is employed in light houses, with the addition of a parabolic reflector, which multiplies the intensity.

The ALCOHOL BLOW PIPE, as represented, is an ingenious and useful instrument. The alcohol in the brass globe, is boiled by the heat of the flame of the lamp placed underneath; this causes expansion or formation into vapour, which issuing out at the jet, is ingeniously made to pass through the same flame, whereby, from its inflammibility, a stream of most intense heat is made to play upon any substance submitted to its influence.

The OXYHYDROGEN BLOW-PIPE is formed on the same principle as the oxyhydrogen microscope. The intense heat produced by the combination of the gasses, when a jet of flame is poured upon any substance, is sufficient to decompose them.

ILLUSTRATION V.

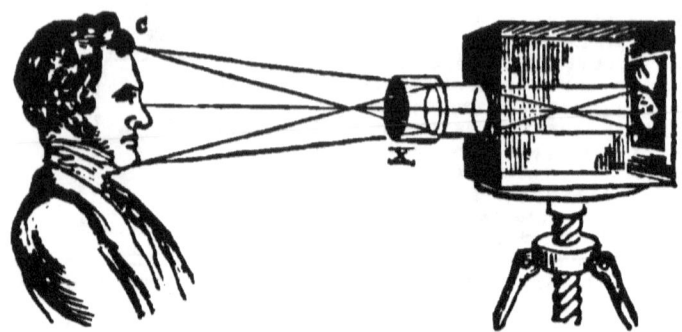

DAGUERREOTYPE.

The sensitiveness of the nitrate of silver to the action of light, has led to the discovery about to be described, which after the name of its inventor, M. Daguerre, is called Daguerreotype. This invention enables us to obtain the perfect representation of a landscape, a picture, a piece of sculpture, or a portrait, almost instantaneously, and so rapid is the progress, that the resemblance of persons in the act of smiling, have been secured by the unrivalled exactness of M. Claudet, at the Royal Adelaide Gallery, who has greatly improved the original invention.

The way in which landscapes and portraits are taken, has been kindly furnished to the author by M. Claudet, and is as follows. A plate is made of pure silver upon copper, which is rendered sensitive to the action

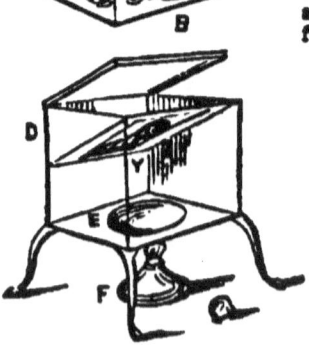

of light, by being suspended, for a short time, over a solution of bromine (B,) the vapour of which combines with the surface of the silver. This plate so prepared is placed within the camera, as seen in the cut A, and immediately that the cover is removed for the light to pass within it, the image of the person is fixed; but this effect is not apparent, nor is the image on the plate visible, until it has been carefully removed (without exposure to the light) from the camera to a box D, in the bottom of which is an iron cup containing mercury or quicksilver, E, the vapour from which, driven off by a spirit lamp, F, is attracted by those parts of the plate which have been affected by the light, and by mixing with the decomposed matter, form a new crystalline compound, which causes the picture to appear on the surface of the plate. After the plate has been thus submitted to the vapour of mercury, until the design has become perfected, it is immersed in a solution of soda, which immediately dissolves and washes away the iodine and other chemical substances from the surface of the picture. Gold is afterwards precipitated upon the design, which fixes and renders it impervious to light and perfectly permanent. Apparatus, enabling any one to take similar portraits and pictures, may be had of Mr. Clarke, optician, Strand, and which having constantly used ourselves, we can confidently recommend.

ELECTROTYPE.

The Electrotype is commonly designated as the mode of obtaining the copy of *coins, medals*, engraved plates, and other objects, by the agency of galvanism; which may be illustrated by the following experiments.

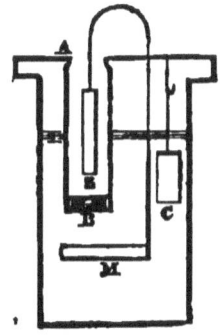

Take an earthen jar and a porous tube, fill the tube with ten parts water and one of sulphuric acid, put it in the jar, into which pour as much dissolved sulphate of copper as will fill three parts of it. Place in the tube a piece of zinc, to which a copper wire is soldered and bent round, so that one end be immersed in the sulphate of copper, and a deposit of the copper will immediately be formed upon the wire. If there be plenty of acid and water, so as to allow of the action enduring for a long time, this process will go on till it has deposited all the copper. This is the principle upon which electrotyping proceeds; a principle referrible to chemical attraction.

The diagram consists of an outer vessel of wood, glass, or earthenware, capable of holding a pint of liquid, within which there is suspended a short lamp glass A, the one end being open and B, the other, closed, with a diaphragm of plaster of Paris; C is a small bag of the crystals of sulphate of copper, to keep up the strength of the solution; Z is the zinc, and M the metal.

To OBTAIN THE COPY OF A COIN OR MEDAL.—Cover the medal with a coat of wax on that side on which the deposited copper is not required to attach itself. The medal thus coated is to be placed in a frame of wire, or if it has no figure on the reverse side, a wire may be soldered to it. To the wire and medal thus arranged, a piece of sheet or cast zinc must be soldered, and we are at once furnished with the materials for the battery, as the object to be copied supplies the place of the copper. Let the apparatus be now charged with the solution, by pouring into the outer vessel a portion of the cupreous solution, so that it will stand about an inch above the medal, then pour in the glass the dilute acid, to the same height as the former; now introduce the zinc into the acid, and the object to be copied into the solution of copper, which will immediately be deposited on the medal, and when of a sufficient thickness may be taken off. An intaglio being thus obtained, the same process is applied to this cast, and a *fac simile* of the original is obtained.

TABLE OF SPECIFIC GRAVITIES.

Distilled water	1,000	Emerald of Peru	8,775
Sea water	1,026	Diamond	3,521
Platinum, purified		Rock crystal	2,650
and fused	19,500	Agate	2,590
Ditto hammered	20,377	Onyx	2,376
Pure gold	19,258	Muscovy talc	2,792
Standard gold	17,486	Common slate	2,672
Mercury at 40°	15,545	Alabaster	2,730
——— at 47°	13,545	White marble.	2,716
Pure silver, cast	10,474	Limestone, from 1,336 to 2,390	
Lead, fused	11,352	Ponderous spar	4,474
Bismuth	9,823	Fluor spar	3,080
Nickel	8,660	Pumice stone	0,914
Brass, cast	8,396	Glass, from 2,520 to 3,437	
Cobalt	7,812	Dry Ivory	1,825
Copper, fused	7,788	Sulphur	1,990
Tin, fused	7,291	Phosphorus	1,714
Iron, cast	7,207	Ebony	1,117
Ditto, bar	7,788	Yellow amber	1,078
Steel	7,846	Spirits of wine	0,837
Zinc	7,191	Concentrated sulphuric acid	2,125
Manganese	6,850	Ditto nitric ditto	1,580
Antimony	6,702	Ditto muriatic ditto	1,194
Tungsten	6,678	Ditto fluoric ditto	1,500
Tellurium	6,115	Oil of olives	0,915
Molybdenum	6,000	Camphor	0,989
Arsenic	5,763	Spermaceti	0,943
Zircon	4,300	Yellow wax	0,965
Barytes	4,200	White wax	0,869
Strontian	3,700	Tallow	0,942
Corunda	3,000	Dry oak	0,925
Silex	2,660	Ash	0,800
Magnesia	2,300	Maple	0,755
Lime	2,300	Elm	0,600
Alumen	2,000	Fir	0,550
Oriental ruby	4,283	Cork	0,240
Garnet	1,118		

N.B. Metals hammered or drawn into wires have an increased specific gravity.

QUESTIONS.

The Laws of Matter and Motion.

CHAPTER I.

1. From what is the term Natural Philosophy derived—What does it *teach*—What does it enable us to *explain*—What to *trace*—What to *ascertain*?   2. Of what does the universe consist—What do we infer concerning it?   3, 4, 5. What is *space*—What is the difference between *absolute* and *relative* space?   6. What is matter?   7. What is it composed of?   8. Name the *primitive forms*?   9. What kinds of matter are *visible*—What kinds are *invisible*?   10. What power causes atoms to unite—What power causes atoms to separate from each other?   11. Of what then is the universe composed?   12. What are the three general states of matter—Describe the difference of these states?   13. What do certain modifications of attraction and repulsion produce?   14. Are the states and substance referred to, organic or inorganic—What is an organized body—What is the organ of sight? &c.

CHAPTER II.

15. What must every existence have—Who is the Great First Cause of all things?   16. What are secondary causes?   17. What advantage is there in our knowledge of causes?   18. Also of effects?   19. Why is natural philosophy, peculiarly the science of cause and effect?   20. To ascertain causes and effects, how does the philosopher proceed—What is the advantage of a mind well stored with facts?

CHAPTER III.

What are the chief qualities of matter?   21. What is solidity?   22. Give an example.   23. Give an example of substances when mixed, taking up less room than when separate?   25. Give an example of bodies when heated, taking up more room.   26. Name some dense and porous bodies.   27. What is the common standard of density?   28. What is divisibility?   29. Why is matter said to be *infinitely divisible*?   30. Give some instances of the divisibility of matter.   31. From the microscope.   32. Lycoperdon.   33. Musk.   34. Gold beaters.   35 The dissolution of a metal in an acid.   36. Wool, silk, spider's web.   37. What is *mobility*?   38. What is *inertia*?   40. Give an example of the ten dency of a body to remain in the same state as regards *rest*—Card and coin—Coin and glasses. 42. Give an example of the tendency of bodies to remain in the same state into which they may be put, as regards *motion*—Dog and hare—What may we learn from these matters?

CHAPTER IV.

43. What do certain modifications of attraction and repulsion produce?   44. On what does the *hardness* of bodies depend—Is it according to their densities?   45. What is *elasticity?*   46. Give examples of elasticity—From vegetables.   47. From minerals.   48. How formed—What proves an ivory ball elastic?   49. To what bodies does *brittleness* belong?   50. From what does *malleability* in bodies arise?   51. What is to be understood by *ductility*—What metals are not ductile?   52. What other substances?   53. Give examples of the *tenacity* of bodies?   54. Repeat the comparative tenacity of some bodies—How much more tenacious is *cast steel* than *gold*—*copper* than *oak*—Give an instance of the tenacity of *iron*.

CHAPTER V.

MOTION—55. What does motion signify?   56. Describe the two kinds of motion.   57. What do the words force and power denote?   58. Name the qualities of motion?   59. What is momentum—Its law?   60. Give an example?   61. What axioms flow from these principles?   62. What is the *first* law of motion?   63. The second?   64. The *third?*   65. Give an example of the first law.   66 Give an example of the second law of motion.   67. Of the *third*.   69. In the consideration of motion, what circumstances are to be attended to, (1, 2, 3, 4, 5, 6)?   70. What is simple motion—What is compound motion?   71. What is accelerated motion?   72. What is retarded motion?   73. What is called the momentum of a body?   74. Give an example.   75. In the case of hard bodies striking each other, how may they be regarded?   77. What happens when any elastic body strikes another?   78. What happens when the difference of size is very great?   79 What does these phenomena seem to constitute?

80. TIME—What does the consideration of motion involve?   81. What is *absolute* time?   82. What is *relative* time?   83. Why is time conceived as measurable?   84. How is it measured?   85. What are its divisions?

CHAPTER VI.

ATTRACTION—86. What is to be uncerstood by attraction?   88. What are the various kinds of attraction, (1, 2, 3, 4, 5, 6)?   89. What is the attraction of gravitation? 90. What is the consequence of this power? 91. Give some instances of the universality of gravity. 92. In what proportion does the attractive force of the earth increase *above* and *below* its surface? 93. Give an example. 94. What is the rate at which bodies fall? 95. Is the force of gravity equal on all parts of the globe—How is it proved not to be equal?

CHAPTER VII.

GRAVITY—96. What is the centre of gravity in a body? 97. Upon what will the method of finding the centre of gravity depend— How found in a cylindrical *rod*—How in a *circular* flat body—How in a sphere? 98. Where will the centre of gravity be in a cylinder— Where will it be found in a cone? 99. How shall we find the centre of gravity in a board? 100. What sometimes happens from want of attention to the centre of gravity—How is attention to it shewn in rope dancing? 101. What are the requisites to enable a body to stand *firmly*—What renders it easily overthrown? 102. Give an example from bodies on an inclined plane. 103. Give examples of instinctive attention to this particular in our common motions.

CENTRAL FORCES—105. What actions or motions of bodies bear the name of central forces? 106. Explain the centrifugal force. 107. Also the centripetal. 108. Give a practical illustration of the centrifugal force. 109. Explain its action in the machinery called the governor. 110. Its effects by rotation of the earth. 111. Advantage of a knowledge of it.

Mechanics.

CHAPTER VIII.

1. What is mechanics? 2. How are the mechanical agents employed? 3. When mechanical agents are combined, what is the instrument formed by this combination called? 4. Name the mechanical powers (1, 2, 3, 4, 5, 6)—What are the first three—What are the last three? 5. What is the *lever*—Where is the *fulcrum*—Where the *weight*? 6. What three things are to be attended to in the lever? 7. From what does the advantage result in this power? 8. Give an example of this. 9. *Power* being gained by this instrument, what is *lost*? 10. Describe the *first* kind of lever—The *second* kind —The *third* kind. 11. Give examples of levers of the *first*, *second*, and *third* kind—How do you find what power will balance any given weight—State the question in an example? 12. How are levers combined—Explain the diagram?

CHAPTER IX.

19. What is the WHEEL and AXLE—Explain its action? 20. Why may it be considered as a lever—What constitutes an increased power of this machine—Give an example in which the power is applied to the axis of the wheel instead of the circumference? 21. In what is the advantage gained by this machine in proportion to? 22. Give an example—What is the wheel and axle in principle—Explain this from the diagram? 23. To what purpose is the wheel and axle applied? 25. Explain its use in the watch from the diagram? 30. Besides being connected by teeth, in what other way may wheels be made to act upon each other—Give an example? 31. The PULLEY—What is the pulley—What does the wheel of the pulley move round—When fitted on a block, what is the wheel then called? 32. How many kinds of pulley are there—What is the advantage of the *fixed* pulley—What of the *moveable?* 34. Where may the advantage of a pulley in changing the direction of a power be seen? 35. Give an example of the use of the simple fixed pulley. 37. How are pullies combined—Upon what principle do they afford support to bodies, and establish an equilibrium—What occasions disadvantage in their multiplied use? 39. What is the advantage of the pulley?

CHAPTER X.

40. To what may the next three mechanical powers be referred? 41. What is an *inclined plane?* 42. Give an example from the diagram. 43. An example from Architecture. 44. An example from a horse drawing a load up hill—In a flight of stairs? 47. What power will keep a body from rolling down, when a plane is inclined to the horizon one third of its whole length—If the height of the plane be equal to half its length? 49. How is an advantage gained by this mechanical power?
50. What is a WEDGE—What is its advantage derived from—How is its power increased—Can the effect of the blow given to drive in the wedge be calculated? 52. What work can be performed by the wedge? 53. What tools are wedge shaped—What animals.
55. What is the SCREW—How do you prove it to be an inclined plane? 56. Name the different parts of a screw? 57. What is the advantage gained by it? 60. How and where are screws generally used? 61. What is a perpetual screw?

CHAPTER XI.

RESISTANCE—62. What resistances are met with in the motions of bodies? 63. What contrivances counteract *friction*—What is the resistance of water—Why does resistance increase in a resisting medium in air, when a body descends in it? 65. Describe the two kinds of friction. 66. What is the friction by rubbing occasioned by—Why is the friction greater between substances of the *same* kind, than between those of *different* kinds—What is the consequence of this? 67. How is the friction of the same substances measured—What is the angle of repose—What is the force required to overcome the friction of *metals* equal to—What for *wood*? 68. What is the friction of the lever—Wheel and axle, pulley, wedge, and screw—What is generally allowed for friction in mechanical contrivances?

FLY WHEEL—69. What is a fly wheel—What is its use—Where may its effects be witnessed—What part of a clock acts as a fly wheel? 70. What irregularities of motion in a machine does a fly wheel overcome—What are the causes of a want of uniformity in the working of a machine, (1, 2, 3)? 71. How are these irregularities equalized by the fly wheel? 72. What does the use of the fly resemble? 73. How is the power of the fly wheel to resist acceleration, determined? 74. How is its power manifested in the turning of a winch? 75. What other property has the fly wheel besides that of regulating the motion of machinery? 76. Give an example of its efficacy in this condition. 77. When used as a regulator of force, where should the fly wheel be placed—Where placed when used as a magazine of power—Why does the fly wheel never retain all the power communicated to it by the first mover?

CHAPTER XII.

PENDULUM—79. What is the pendulum—How are its vibrations produced? 80. What was Galileo's discovery—How is this proved by example? 81. What is the length of a pendulum vibrating seconds in the latitude of London? 82. Why must a pendulum to vibrate seconds, be shorter at the equator than at the poles? 83. What inventions are those which correct the contraction and expansion of pendulums? 84. What is the centre of oscillation—How do you prove that the time of a pendulum's vibration, is not affected by varying its weight—How do you prove it is not affected by the distance which it swings? 85. What has the pendulum been considered as a proper basis for—who determined its real length—To what is it principally applied? 86. What substance would be most suitable for an attached pendulum? 87. What is the first essential in the construction of a clock, and why? 88. In what degree does a brass rod of the pendulum length, expand or contract, in the variation of one degree of temperature—What effect will this have upon a clock? 89. Describe the mercurial pendulum of Graham—How is the amount of compensation determined? 90. Describe the gridiron pendulum and its application.

Hydrostatics.

CHAPTER XIII.

1. What does *Hydrostatics* treat of—What is the difference between this science and *Hydraulics*? 2. Name the non-elastic fluids—What are the elastic and compressible—What science refers to these? 3. What is the difference between the pressure of a solid and a fluid? 4. What is the ratio of proportion of the pressure of fluids on vessels—Explain by diagrams? 6. What is the hydrostatical paradox? 7. Give an example from the diagram. 8. Explain the upward pressure of fluids, by a description of the hydrostatic bellows. 10. To what principle is the hydrostatic screw to be referred—Explain the construction of this machine? 11. What is the rule for estimating the pressure of fluids upon horizontal surfaces? 12. What is the rule for surfaces *not* horizontal? 13. How do you find the pressure on a dam or flood gate? 14. How upon the sloping *side* of a pond? 15. How is the pressure of a fluid against the upright sides of a cylinder found? 16. Mention the laws concerning the pressure of fluids upon solids immersed in them.

CHAPTER XIV.

SPECIFIC GRAVITY—17. What is the specific gravity of a body? 18. What is the standard of this gravity? 19. Relate how the method of ascertaining specific gravities was discovered by *Archimedes*? 22. What instrument is used for ascertaining the specific gravity of solid bodies—Describe this instrument, and give an example? 25. What are the methods employed for finding the specific gravity of solids *lighter* than water? 27. Give an example. 28. Give the principle upon which the specific gravities of fluids are ascertained. 29. How found by the hydrometer—State the propositions upon which the use of this instrument depends? 29. Which of these propositions gave rise to the graduated scale hydrometer? 30. Describe this instrument, and the manner in which it is used. 31. Describe the hydrometer with weights, and the way in which it is used. 33. What is the use of the hydrometer? 34. Explain the method of finding the specific gravity of a fluid or of an aeriform body, by the 1,000 grain bottle—How is the weight of any given bulk of a body calculated, when its specific gravity is known (some other examples should be here given to the pupil from the table of specific gravities which will be found in the illustrations at the last page.)

Hydraulics.

CHAPTER XV.

1. What does the science of hydraulics comprehend? 2. How only can water be set in motion? 3. Give an example of water always finding its level, and an example of the application of this law. 5. Explain the principle upon which water rises through bended pipes and fountains, and state why the water does not rise quite to its level head in these modes of conveyance? 6. What is the law of the horizontal force of water flowing from a vessel—State the reasoning upon this, after the word *because* in the text? 7. Explain the diagram. 8. What is the velocity with which water would issue from a given depth below the horizontal surface, as at the aperture B.—What is the guide? 9. What is the *ratio of pressure*—Give examples of this? 10. Relate the experiment. 11. Describe the syphon. 12. Explain its action. 13. How are intermitting springs accounted for?

CHAPTER XVI

Pumps.—14. Who is supposed to have been the inventor of the pump? 15. How many kinds of pumps are there? 16. Describe the action of the sucking pump. 18. Describe the lifting pump and its mode of action—Describe the action of the fire engine (see illustration, page 252.) 23. What was the most ancient hydraulic machinery? 24. Describe the *Persian wheel*. 25. Explain the *chain pump*. 26 The *screw of Archimedes*. 26. The centrifugal pump. 27. Explain the action of Baker's mill. 28. What is the difference between an undershot and an overshot water mill?

Pneumatics.

CHAPTER XVII.

1. What is *Pneumatics?* 2. What is the air? 3. What is the height of the atmosphere? 4. The air being a transparent body and invisible, how do we discover its presence—Why cannot we see the air? 5. What are the *qualities* of the air? 6. What is the colour of the air, and how is its apparently blue colour to be accounted for? 7. How does the weight and pressure of the air exhibit itself—How is its pressure demonstrated? 8. What are the phenomena resulting from atmospheric pressure, (give examples from paragraphs 9, 10, 11, 12)? 13. What is the estimated aggregate pressure of the air upon the surface of the globe? 14. What is this pressure equal to on an inch of surface—What would be the height of a column of mercury an inch square, to equal this weight—What

would be the height of a column of water an inch square, to equal it? 15. What is the pressure of the air on the body of a man equal to—Why does not this pressure destroy him? 16. What results from the weight of the atmosphere being equal to that of a body of water, 33 feet high? 17. Upon what principle did Torricelli invent the barometer? 18. Explain the principle and the action of the common pump?

CHAPTER XVIII.

19. Is air always of the same density—To what is its density proportionate—Why is it more dense at the earth's surface, than in the upper regions—In what proportion does the mercury descend as a person *ascends?* 20. Describe Guy Lussac's experiment—What did this prove—What is the term for expressing the lightness or thinness of the air in the upper regions? 21. How may the different densities of the air be illustrated? 22. How may the *elasticity* and *compressibility* of the air be shown? 23. Give other examples, (sugar; glass blowers). 24. What were the results of Boyle's experiments, relating to the expansion of air? 25. What is the consequence of the principle of the air's elasticity—What would be the density of the air 50 miles below the surface, (explain the consequence of the particles of the air coming into contact by pressure, by a reference to the match syringe)?—What is the law of the elasticity of the air? 27. How is the materiality, elasticity, compressibility, and pressure of the air proved by the diagram? 28. Into what space may air be condensed? 29. Describe the condenser and its mode of application. 30. How does atmospheric pressure change the temperature at various altitudes? 31. Explain the *air gun.* 33. The *wind gun.*

CHAPTER XIX.

AIR PUMP—35. What are the principles upon which the air pump acts—Who was its inventor and improver? 36. What happens under the exhausted receiver of the air pump? 37. Explain the construction of the pump. 39. What experiment proves the *downward pressure* of the air? 40. What experiment proves that the air has *weight?* 41. How do you find the weight of air contained in a room? 42. What experiment proves the specific gravity of the air, compared with *that of water?* 43. What experiment proves the air to be elastic? 44. What experiments prove the resistance of the air? 49. What experiments prove the *pressure* of the air—Explain the methods of performing the experiments related in paragraphs 50, 51, 52, 53, 54, 55, 56, 57, 58, 59, 60, 61, 62, 63.

QUESTIONS.

CHAPTER XX.

BAROMETER—64. What is the barometer—What does it consist of—How is it made—Explain the principle of its construction? 65. What is the *standard altitude*—What is the *scale* of variation? 66. What is the air pump barometer, and what does it *indicate*—Describe the construction of the wheel barometer—Give the rules for predicting the state of the weather from paragraphs 68, 69, 70, 71, 72, 73, 74, 75, 76, 77, 78—How is the barometer employed in measuring heights?

THERMOMETER—80. What is the thermometer—Explain its construction? 81. What are the thermometrical scales in general use? 82. How are degrees of Fahrenheit's thermometer converted into those of Reaumer, &c.—At what degrees does water freeze in the several scales? 85. What is the hygrometer—Describe the weather house? 86. Describe the catgut hygrometer. 87. Also the pulley hygrometer. 88. The balance hygrometer, &c.

CHAPTER XXI.

ÆRONAUTICS—1. What is the science of æronautics? 2. What is the air balloon—Why does it ascend in the atmosphere? 3. Explain the proportion of the expansion of heated air? 4. What is a fire balloon, and by whom was it first invented? 5. What are the materials of modern balloons, and with what are they inflated? 6. How are the superfices of a balloon computed? 7. Name some of the principal æronauts? 8. To what is the extraordinary velocity of balloons to be ascribed? 9. Describe the Frenchman's experiment.

The DIVING BELL.—10. How can the principle of this machine be illustrated? 11. Explain the diving bell—What was Smeaton's diving bell—Describe that of Rennie—Explain the diagram of the diving bell—Describe the dress of the submarine diver, &c?

Meteorology.

CHAPTER XXII.

1. What does *meteorology* treat of? 2. What is a *meteor*, and of what is it composed? 3. Why does hydrogen gas rise in the atmosphere—What does the mechanical and chemical action going on in the atmosphere produce—How is the electrical equilibrium restored? 4. Describe the process of natural evaporation. 5. How is rain produced? 6. What is *snow* and *hail*—How produced? 7. How is *dew* produced? 8. Further explain the economy of nature in evaporation. 9. What are *fogs* and *clouds?* 10. What are *mists?* 11. How is *wind* produced? 12. What influences the motion of the atmosphere in and near the equator? 13. What are *constant, perio-*

dical, and *variable* winds? 14. What is the region of the *constant* winds? 15. Where do the *periodical* winds prevail? 16. What is the cause of the sea and land breezes in tropical latitudes? 17. Describe an experiment relating to this phenomenon. 18. What wind generally prevails in England during the Spring—What is supposed to be the principal cause of the variableness of wind in our country? 19. In what ratio does the force of the wind increase? 20. What is its estimated force in pounds on a square foot? 21. What is the estimated pressure of the breeze, gale, and hurricane? 22. What are *hurricanes?* 23. What are *tornadoes?* 24. What are *whirlwinds?* 25. Describe the formation and course of a water spout—To what causes are whirlwinds and water spouts attributed? 26. Define the seven classes of clouds—What takes place in clouds when *rain* is formed? 28. Where is the *cirrus* seen, and what does it portend—What is the shape of the *cumulus*—Describe the appearance of the *cirro stratus*—Of what are clouds composed?

Acoustics.

CHAPTER XXIII

1. What is the term *acoustics* derived from—What does the science treat of? 2. What produces *sound?* 3. When these impulses are quickly repeated, to what do they give rise—What does the tone or pitch of this *continued sound* depend upon? 4. Give an illustration of the formation of this continued sound and tone. 5. When a continued sound is produced by impulses which do not fall in regular succession, what are they said to produce? 6. Explain how sound arises from vibration. 7. How is sound propagated? 8. What is all sound assumed to be? 9. What is the effect of sounds in liquids and in solids—Give instances of this fact? 10. Describe *Biot's* experiment. 11. Describe the stethoscope, and the principle upon which it acts. 12. Give some illustrations of the way in which sound is *communicated*.

The EAR.—13 and 14. What are the principal divisions of the *ear*— What is the external part called? 15. After the vibrations of sound are collected by the *concha*, on what do they fall? 16. What does the second division of the ear consist of—Why is the *tympanum* so called—What forms the communication between this membrane and the circular opening called the *foramen ovale?* 17. Name these bones in their order and explain their parts. 18. What is the *third* division of the ear—Of what does it consist—Where does the acoustic nerve terminate—Describe the action of the different parts of the ear?

CHAPTER XXIV.

19 Explain the law of the angles of incidence and reflection, as regards sound. - 20. When are echoes distinguished? 21. What is extraordinary in the natural echo of the Rhine? 22. Give an instance of the effects of the *concentration* of sound by concave surfaces —Explain the diagram? 23. Give some illustrations of the *reflection* of sound from concave surfaces. 25. What contrivance in relation to these principles, produced the exhibition called the invisible girl.

CHAPTER XXV.

26. Upon what do the sounds called *bass, low*, or *grave* notes depend? 27. Upon what does the *frequency* of vibrations on strings depend—Explain the reason? 28? What is the cause of *accordant* and *discordant* notes? 29. Give an instance in proof of this—To comprehend the nature of *protracted vibration* or *resonance*, to what must we advert—Why does this afford a correct explanation of the phenomena of *resonance?* 30. What is the difference between *melody* and *harmony?* 32. What is the principle upon which the *tuning fork* is employed? 33. Repeat the experiment relating to the *reciprocity* of sound? 34. Describe the musical glasses of Dr. Arnot, and the advantage of their arrangement? 35. What is the result of the vibration of rods, and its practical application? 36. Describe the *Kaleidophone?* 37. In what does the vibration of *plates* differ from those of *rods*—How have these vibrations been traced through their variations—What other methods have been employed for shewing the nature of vibrations? 38. What are the phenomena of the *vibrations* of a *cord?* 39. Into what distinct classes may musical instruments be divided? 40. Explain the action of *stringed* or membraneous instruments. 41. How is *harmony* occasioned in the clarionet, flute, &c.—How in the French horn, serpent, &c.—Upon what are metallic instruments dependent for their *diversity* of tone? 42. What principle seems to prevail throughout these instances? 43. Explain the manner in which the *pitch* or *tone* of a wind instrument is produced. 44. What is the compass of the harp, piano, guitar, clarionet, horn, bazoon, flute, violin, violincello, human voice, soprano, tenor, &c.

Optics.

CHAPTER XXVI.

1. What is *optics*—How is it divided? 2. What are the theories concerning light? 3. What is the *Newtonian* theory—What is the *undulatory* theory of *Huygens*—What is light—What is its absence

called—When it passes through bodies, what are they called—When it enters these bodies obliquely, what happens to it—Does every part of a beam of light, passing through these bodies, bend equally—What happens to it? 5. What law does light follow? 6. What does light consist of—How are these rays perfected—How is this proved—What is every quantity of light composed of—What is a beam of light—What is a pencil of rays—What is the point at which rays meet, called? 7. What is the *velocity* of light—How was this velocity discovered, and by whom? 8. Of what excitements is light the effect of? 9. What are those bodies called which *emit* light—What are those called which *transmit* it—What are those called that reflect part—What are those called which do not suffer it to pass through them?

CHAPTER XXVII.

Dioptrics—What is dioptrics—Explain the law of refraction from the diagram? 11. What happens when a ray of light passes from glass to air? 12. How is the refraction of light shown by the *prism?* 13. Explain the diagram of the coin in the basin of water. 14. To what important axiom do these facts lead—Give an illustration? 15. What are the refractive powers of media dependent upon? 16. From what did Newton infer the combustibility of the diamond? 17. How does the multiplying glass illustrate the theory of refraction? 18. What phenomena are observed relating to this subject, on a day of bright sunshine? 19. To what is the mirage referrible.

CHAPTER XXVIII.

Catoptrics—What does the word catoptrics signify, and of what does the science treat? 21. Give an illustration? 23. What axiom relates to this experiment? 24. Apply this rule to plane surfaces, and explain the diagram? 25. What is the reason of the double image of a lighted candle in a looking glass? 26. Explain the laws of refraction from a *convex* and a *concave* surface? 27. Explain the law of the rays of light, when falling on a concave mirror? 29. What is the focal distance of all kinds of mirrors—From what is the word *focus* derived—Explain the illustration as regards heat? 31. Describe the experiment of the bottle of water and concave mirror 32. Explain the cause of the appearance of the image in the air between the mirror and the object. 33. For what are concave mirrors principally used—How do they act, and what are the phenomena connected with them?

CHAPTERS XXIX, XXX, XXXI.

Prisms, Lenses—34. What is a lens? 35. Describe the forms of the various lenses. 36. What are *converging* and *diverging* rays? 37. When parallel rays fall on a *plano convex* lens, where do they meet? 38. Where do they meet when they pass through a *double convex* lens—In what proportion is the power of the heat collected in the focus (f) of the diagram? 40. What degree of divergence takes place when parallel rays fall on a double *concave* lens —Explain the diagram? 43. What happens if rays come more convergent to such a glass? 44. When the object is within the focus of a concave or convex lens, what is the position—Why are the images imaginary—Why are the images of objects placed *beyond* the focus of a convex lens *inverted* and *real?* 45. What are the elementary parts of a ray of light? 46. How is a ray of light decomposed? 47. Which is the least refrangible of these rays, and where is its position? 48. How can white be produced by the union of these colours? 49. How is the decomposition of light by absorption produced—What happens in this experiment? 50. What conclusions were drawn from these phenomena—What is the general conclusion? 51. What did Sir William Herschel discover with regard to the heating powers of the spectrum—Who confirmed his results, and what measures were obtained—Where is the place of maximum heat? 52. What did M. *Fraunhofer* discover by means of a *photometer*—What else did that philosopher discover—What is the most important practical results of these fixed lines? 53. What are the chemical effects of the spectrum? 54. What is the most remarkable effect of solar light? 55. From what does the *colour* of bodies arise—What are the transformations of *visual* colours—What colours absorb the most odours? 56. What are the degrees of heat in a prismatic spectrum —What does the different colours of very remote stars prove—From what do the *green, red, indigo,* and *violet* colour of plants arise—What are the heating powers of the red rays in the spectrum, to *green* and to *violet?*

CHAPTER XXXII.

The RAINBOW.—57. What does this meteor in the heavens depend upon—Explain the formation of the primary bow? 61. Why does not the rainbow immediately disappear upon the rain falling to the ground? 62. Explain the formation of the secondary bow. 63. What are the numbers of refractions and reflection in the primary bow—What are the numbers in the secondary? 64. What is the angle made by the violet rays with the incident ones—What is that of the red rays—which bow is formed by fewest rays.

CHAPTER XXXIII.

The EYE.—What is the human eye considered optically—Describe its shape—Its orbit—Its situation? 66. What is the use of the *eyebrows?* 67. Of the *eyelids?* 68. Describe the lachrymal gland. 69. What is the remedy in disease of this gland? 70. How are the eyelids stiffened? 71. What are the uses of the eyelashes? 72. What means are provided for giving moisture to the eye—Describe the oblique, called the trochlea, muscle? 73. What are the two first coats of the eye, and how are they disposed? 74. What is the iris—Describe the action of its muscular organization—What is the situation of the *pigmentum nigrum?* 75. What is the third or inner coat of the eye called—What is its situation—Where does it terminate? 76. What is the use of the retina—What is its colour—Why does it appear black? 77. What are the names of the three humours of the eye? 78. What are the humours of the eye intended for—Describe the shape and situation of these humours? 79. How are these lenses adjusted to their proper focus—What would happen without this beautiful contrivance? 80. What experiment will exhibit the manner in which the image of any object is painted on the retina? 82. Explain the mechanism of vision from the diagram. 83. Which is the most powerful of the lenses of the eye, and why? 84. Why do we see things in their natural position? 85. What is one of the most wonderful circumstances relating to the eye? 86. What are we called upon to admire in the construction of the eye—What other remarkable facts bear out this remark? 87. What is the camera obscura? 88. How is it made, &c.? 89. Describe its action. 90. Describe its form when used as a public exhibition. 91. Describe the camera lucida. 92. Describe the magic lanthorn. 93. What is the difference between this and the camera obscura?

Astronomy.

CHAPTER XXXIV.

1. What is *astronomy?* 2. How is it divided—What is the difference between *pure,* or *plane,* and *physical* astronomy? 3. From what people do we trace the earliest history of astronomy—Explain the Egyptian system? 4. Who among the Greeks cultivated this science? 5. What was the state of astronomical knowledge in the first century of the Chrisitan era? 6. Who cultivated the science among the Arabians? 7. Who revived the system of Pythagoras—What was the system of Tycho Brahe—What did Keplar ascertain? 8. What resulted from the invention of telescopes? 10. What were the discoveries of *Helvetius, Huygens, Newton* and *Herschel.*

CHAPTER XXXV.

11. Describe the solar system? 12. Give the order of *the* planets —What does this system also include—What else does it assume?

13. Explain the centrifugal and centripetal forces? 14. What is the shape of the planetary bodies? 15. Explain the inclination of the planetary orbits—What is the ecliptic and its plane—What are the ascending and descending nodes—Name the signs of the ecliptic? 16. Describe the form of a planetary orbit—Which are the foci—When is the sun said to be in its *aphelion* and *perehelion*—At what point is he said to be at his mean distance? 17. What is the line of the *apsides*—What is the lower or congregate axis—What is the eccentricity of the orbit—Where are the higher and lower foci—Where is the lower apsis—Where the higher apsis.

The Sun.—What is the shape and diameter of the sun—What is the time of his rotation—What is his *distance* from the earth—How much *larger*—What is the extent of his surface—What is his density—Of what is it supposed he consists? 19. What are his *real* motions? 20. What are his *apparent* motions? 21. What are the spots seen on his surface?

CHAPTER XXXVI.

Mercury.—Give his distance—Motion in orbit—Annual revolution. 23. What is his aspect, and when is he seen? 24. What is his size, and what does he exhibit? 25. Describe his transit? 26. What are the laws of his motion—What is his density—What is diurnal revolution?

Venus.—What is the appearance of Venus, and her daily rotation? 28 What is the inclination of her axis? 29. What is her diameter—Distance from the sun—Length of her year, and motion per hour? 30. How often does she pass over the face of the sun? 31. When is she a morning, and when an evening star? 32. What did M. Schroter discover on her surface—What is the length of her day.

CHAPTER XXXVII.

The Earth.—Give her *polar, equitorial*, and *mean* diameter, circumference, and distance from the sun—Describe her shape? 35. Give her annual and diurnal motions—What is the *precession* of the equinoxes? 37. Give the inclination of her axis, and state to what it gives rise—Explain the diagram.

CHAPTER XXXVIII.

The Moon.—39. Give her *shape, diameter, circumference, period of rotation* and *motion* in her orbit? 40. Explain her phases from the diagram? 41. Describe her orbit—Explain the nodes and the terms *apogee* and *perigee?* 42. Describe her telescopic appearance? 43. What is the libration in longitude? 44. What is the libration in latitude—What is the density of her atmosphere? 45. What is the cause of the phenomenon of the harvest moon?

CHAPTER XXXIX.

Mars.—Describe Mars—His *shape*, the *eccentricity* of his orbit. 47. His diameter. 48. Telescopic appearance. 49. What are the asteroids? 50. Name them. 52. What is remarkable about their orbits? 55. What also is remarkable about these planets?

Jupiter.—56. Give his distance—Rotation on axis—Annual revolution—Inclination of axis—And diameter. 59. Describe his belts? 60. The spots on his surface? 61. How many satellites has he, and give the period of their revolution? 62. How do we find the longitude by the eclipse of Jupiter's satellites? 63. What is a *transit*— —What is an *occultation?* 64. Give his density and the inclination of the plane of his orbit?

Saturn.—65. What is remarkable about Saturn—Describe his ring —What did Huygens discover? 68. What is the distance between the innermost part of the ring and his body? 69. Do these rings rotate? 70. What else has Saturn belonging to him?

Herschel.—71. What is his diameter and annual revolution? 72. Who discovered him—What is his *distance*, number of *satellites*, *diameter, periodical revolution,* and *inclination of orbit?*

CHAPTER XL.

The Tides.—76. To whom are we indebted for a knowledge of the tides? 77. What are they occasioned by? 78. Give an illustration of them. 80. Explain the diagrams. 82. Upon what other causes do the height of the tides depend?

Eclipses—83. How many kinds of eclipses are there—What is an eclipse of the sun—What is an eclipse of the moon—Explain the diagrams? 84. What is a *total*—a *partial,* and an *annular* eclipse? 85. When can a total eclipse of the sun happen—Why does it not happen at every conjunction? 86. What is the penumbra, and how is it occasioned? 87. How is the quantity of an eclipse expressed? 88. What number of eclipses can or may happen annually?

CHAPTER XLI.

Stars.—89. Why called *fixed*—What are they supposed to be— Distance of the nearest—How known? 90. Are they really fixed? 91. How are stars divided? 92. Why formed into constellations— Number of constellations? 93. What are periodical stars—Have any new stars appeared? 94. What is the milky way? 95 What are nebulæ, and where found? 96. What was Herschel's opinion concerning them? 97. What is the computed distance of the fixed stars? 99. What are comets? 100. Mention some other particulars concerning comets.

Chemistry.

CHAPTER XLII.

1. Define chemistry. 2. With what has it a connection? 3. How does chemistry make us acquainted with the internal structure of bodies—Explain decomposition and combustion? 4. Upon what principles do these two methods of induction depend? 5. Explain observation, experiment, and analogy, and say to what they lead. 6. Explain the difference between a chemical and mechanical change. 7 State what mechanical action is attended with, and what takes place in a chemical process. 8. Give an example of chemical *combination* 9. Give an example of chemical *decomposition*. 10. State and explain the *difference* between a mechanical and a chemical *combination*. 11. What *change of properties* attends chemical combination? 12. Give examples of these changes. 13, 14. State the effects of decomposition.

CHAPTER XLIII.

15. What is the number of elementary substances? 16. How are they arranged. (See Table.)

CHAPTER XLIV.

OXYGEN.—18. From what is the word oxygen derived? 19. What is oxygen? 20. How only can it to be obtained—Where does it exist in combination? 21. State its specific gravity, and other particulars relating to it? 22. From what substances can it be disengaged? 23. What effect has the act of respiration upon it? 24. What is its action in germination? 25. What is an oxide?

HYDROGEN.—26. Derivation of the word—The weight and specific gravity? 27. Where is it found, and how procured—What is the ignis fatuus? 28. Can hydrogen be respired—Will it explode—Does it support combustion? 28. How is it made? 30. What is its product when burnt— Describe the process of the *decomposition of water*, and its subsequent *re-formation?* 31. How else can water be decomposed? 32. What is the chemical name for water—State the particulars relating to it? 33. State the name of the principal mineral waters—What is hard, and what is soft water?

NITROGEN.—34. What is nitrogen—How is it obtained? 35. Who discovered it—What is its use in the air? 36. What is nitric oxide? 37. What is nitrous oxide? 38. What is the effect of nitrous oxide when respired?

CHAPTER XLV.

CHLORINE.—39. Describe its properties? 40. How is it made? 41. How used in bleaching? 42. How as a disinfector? 43. What is *iodine?* 44. What is *bromine?* 45. What is *fluorine?*

CHAPTER XLVI.

CARBON.—46. What is carbon? 47. Where is it found? 48. What is charcoal—State its chemical qualities, and its use? 49. What proportion does it form of iron, steel, and plumbago? 50. What is carbonic acid gas, and from what may it be obtained? 51. State the properties of this gas—In what substance is it found in nature 52. Can it be respired—What is its use in the air? 53. What is carburetted hyIrogen—When does it become explosive—How is its explosion prevented? 54. Recapitulate the nature and uses of carbon.

SULPHUR.—55. What is sulphur—Where found—How prepared—Its properties? 56. State the principle properties of sulphur. 57. What is phosphorus—How procured—Its state, properties, and uses?

CHAPTER XLVII.

NON-ALKALINE EARTHS.—58. What are the bases of these earths? 59. Who discovered their metallic nature? 60. What is Silica? 61. What is Alumina—In what does it exist? 62. What are Glucina, Ittria, Zirconia, and Thorina?

ALKALINE EARTHS—63. What are the bases of these? 64. How is lime found? 65. What is mortar? 66. How is lime converted into calcium? 67. What is Baryta, and what are its properties? 68. What is Magnesia—For what is Sulphate of Magnesia an antidote?

CHAPTER XLVIII.

METALS.—74. What distinguishes the metallic bodies? 75. Where are they found—In what do the metals differ from each other —How is this proved? 76. What is the course of metalliferous veins—What is the order of their density? 77. Give the nature and properties of platina, where it is found; and also the same of the other metals. State their *appearance, chemical* and *mechanical properties* —Where *found*—How *used*, &c., including paragraphs from 77 to 99.

CHAPTER XLIX.

ACIDS and SALTS.—100. What are acids, and how do they form salts—State them in their order of classification. 101. State severally the manner in which they are obtained—Their nature—Their effects as chemical agents—How used in the arts—And how they act upon the substances with which they are brought into contact in nature and the arts, from paragraphs 101 to 110.

SALTS.—What is a salt? What are the number of salts? How are they denominated? Explain their nomenclature. 112. Why is Glauber's salt called sulphate of soda, &c.? 113 What is the termination of the names of the salts formed of acids—What denotes a salt to have an *excess* of acid—What denotes it to have a less degree?

TOXICOLOGY.—114. State the antidotes of the most common of the poisons.

CHAPTER L.

HEAT.—115. Where is caloric found—What is heat considered as a sensation? 116. What are the two states of caloric—What is free caloric—What is confined caloric—How do we ascertain that different bodies do not contain an equal quantity of caloric when heated to the same temperature. 118. What is latent caloric—How is the expansion of bodies by heat easily proved—What does the experiment in paragraph 118 prove? What does the experiment in paragraph 119 prove? 120. How do we prove that all bodies do not radiate alike? 121. What is combustion? 122. What are the simple combustibles—What are the compound? 123. What are the supporters of combustion—What takes place during the combustion of a body? 124. What is the state of a body when fully burnt—What is the chemical name given to burnt bodies—What does the process of combustion perform?

Electricity.

CHAPTER LI.

1. From what is the term electricity derived—Upon what does it treat—What produces it—What are electrics? 2. How is the nature of it shown? 3. What do these experiments prove—Name some of the principal conductors. 4. What is insulation—Explain the situation of a person standing on a glass stool when charged with electricity? 5. What happens when an electric as a rod of glass is presented to the ball B in the diagram—What happens if, after being withdrawn, the glass rod be again presented—How can the rod be deprived of its electricity—What happens if we present a piece of sealing-wax to the ball—What do these experiments prove? 6. But what happens if we, after having conveyed electricity to the ball B by means of *excited* glass, as in the first experiment, present to it excited sealing-wax—What happens if the experiment be reversed—What follows from these experiments? 7. What are the laws of opposite electrics?

THE ELECTRICAL MACHINE.—8. Describe its mode of action—State how negative electricity is obtained—Why does the insulated person obtain more than his natural share of electricity? 9. What is the principle of the electrical machine? 10. How will a person have less than his natural share of electricity? 12. What are the theories of electricity? 13. Explain the modes of its excitation and accumulation. 14. What is the cause of its *distribution*—And what is the law of its force? 15. What is the law of its *transference*? 16.

Explain electric attraction and repulsion. 17. What is the Franklin theory of electricity? 18. Explain electrical induction. 19. Describe the Leyden Jar, and its mode of action. 22. The electrical battery—And use of the discharging rod. 23. What is the electrometer—And how does it act? 24. Explain the diagrams of the universal discharger, 25, Du Luc's column, 26, working power of electricity, and, 27, ringing bells. 28. What are the effects of the electric shock? 29. Describe the appearance of the electric spark. 30. How was the identity of lightning with electricity proved? 31. What is the nature of animal electricity? 32. What kind of electricity is evolved by the contact of different metals? 33. In what other ways is electricity excited? 35. Explain the decomposition and reproduction of water by electricity. 37. Explain the phenomena of electrics in a thunder-storm.

Galvanism.

CHAPTER LII.

1. Relate the discovery of galvanism. 2. State its chemical theory. 3. The process adopted for producing it. 4. Describe the voltaic pile. 5. The galvanic trough. 6. What is a galvanic battery? 7. State how it has been improved. 8. Explain the deflagrator. 9. State the electrical and chemical effects of galvanism. 11. Repeat the effects of galvanic batteries. 12. State the physiological effects of galvanism.

Magnetism.

CHAPTER LIII.

1. What is the theory of magnetism? 2. What is the natural magnet? 3. What is polarity? 4. Describe the mariner's compass. 5. What is the azimuth compass? 9. What is the variation of the needle? 10. How do the daily variations occur? 11. What is the dip of the needle? 12. What is animal magnetism? 13. What is electro-magnetism? 14. Why does a galvanic current change the polarity of the needle? 15. How is magnetism communicated to soft iron? 17. How are magnets made by galvanism? 18. What is the effect of a current of electricity upon a magnet? 19. What is thermo-electricity?

Experiments in Natural Philosophy.

MECHANICS.

Solidity.

ILLUST.—If a piece of wood, or stone, occupy a certain space, before you can put another body into that space, you must first remove the stone or wood; and though fluids do not appear at first to offer such resistance, yet, under any circumstances, they will be found to retain this property in an equal degree.

EXP. 1.—Put some water into a tube closed at one end, and insert into it a piston, or a piece of wood or metal, that perfectly fits the inside ! you will find it impossible, by any pressure, to get the piston to the bottom without breaking the tube.

EXP. 2.—If you try the same experiment with the tube empty, as it is called, but in reality filled with air, you will find the same impossibility of putting the piston to the farthest end of the tube.

COROLLARY.—Hence, both water and air, and every other fluid, are equally impenetrable, in this sense of the word, with a piece of marble or steel.

Inertia.

ILLUST.—No one can suppose that matter can begin to move of itself, unless it be in some way acted upon; but it does not appear so evident that it has a tendency to continue in motion for ever. Most people are apt to suppose that all matter has a propensity to fall from a state of motion into a state of rest; because we see all the motions upon the earth gradually decay, and at last totally cease. But this is owing to the resistance of the air, and to friction: for if these be deminished, the body will move longer; and if they could be removed altogether, the body would continue for ever in motion. But take some familiar experiments.

Exp. 1.—A marble shot from the fore finger and thumb, would run but a small distance on a carpet: its motion would be continued much longer on a level pavement; and longer still on fine smooth ice, such as might cover the bosom of the Arar. In this case, the friction is greatest on the carpet, and least on the ice. And if, as we have observed, the friction were entirely destroyed, as also the resistance of the air, the marble once put in motion by a school-boy, would continue in that state for ever.

Motion of the parts of bodies among themselves.

Pages 13—56.

Exp. 1.—Hold a decanter of clear water in the rays of the sun, and you will see that the light particles, which float in the water, are in perpetual motion, which proves that the parts of the water themselves are in constant motion.

Exp. 2.—Let the rays of the sun pass through a small hole in a window shutter, and you will observe that the particles, floating in the atmosphere, and of whose existence you were not before aware, are in constant motion.

Obs. I.—If we reflect a little, we shall discover, that the particles of the most solid bodies are continually changing their situations. Heat expands, and cold contracts, the size of all bodies: now, we know from experience, that the temperature of bodies is constantly varying, consequently, the particles must be in continual agitation, in order to adapt themselves to the size of the body. This is one of the causes of the perpetual motion of the particles of matter, but there are, no doubt, an infinite number of other causes, which escape our observation, or which we are incapable of discovering. The gradual changes that take place in all bodies during a series of years, sufficiently prove that they are constantly acting upon each other; and perhaps there is no particle of matter whatever that is absolutely at rest, but that all are in perpetual action and motion.

Velocity of Motion.

The *Velocity of Motion* is estimated by the time occupied in moving over a certain space, or by the space moved over in a certain time. The less the time, and the greater the space moved over, the greater is the velocity; on the contrary, the greater the time, and the less the space moved over, the less is the velocity.

ILLUST. 1.—To ascertain the degree of velocity, the space run over must be divided by the time.

For example, suppose a body moves over 1,000 yards in 10 minutes, its velocity will be 100 yards per minute. Again, if we would compare the velocity of two bodies A and B, of which A moves over 54 yards in 9 minutes, and B 96 yards in 6 minutes, the velocity of A will be to that of B in the proportion of 6 : 16; for $\frac{54}{9}=6$, and $\frac{96}{6}=16$.

ILLUST. 2.—To measure the space run over, the velocity must be multiplied by the time; for it is evident, that if either the velocity or the time be increased, the space run over will be likewise increased; that is, it will be greater. If the velocity be doubled, then the body will move over twice the space in the same time; or if the time be twice as great, then the space will be doubled: but if both the velocity and time be doubled, then will the space be four times as great.

For example, if two persons set out together on a journey, and one walks two miles and a half, and the other walks five miles in an hour, the velocity of the latter will be double that of the former.

COROL.—It follows from the foregoing reasoning, that when two bodies move over unequal spaces in unequal times, their velocities are to each other, as the quotients arising from dividing the spaces run over by the times. For, if two bodies move over unequal spaces in the same time, their velocities will be in proportion to the spaces passed over. And if two bodies move over equal spaces in unequal times, then their respective velocities will be inversely as the time employed: thus if, A in one second and B in two seconds run over 100 yards, the velocity of A will be to that of B as 2 to 1.

Attraction and Repulsion.

ILLUST.—It is supposed that, besides this attractive force, there is also a sphere of repulsion, that extends to a small distance round bodies, and prevents them from coming into actual contact with each other, except some force be exerted to overcome this repulsion: accordingly, bodies which appear to touch each other, are not in actual contact, and it requires some force to bring them together; and then the attraction of cohesion takes place.

EXP. 1.—If two pieces of lead with flat surfaces be scraped clean with a knife, and squeezed together, they will adhere so firmly, that they can scarcely be separated. The same takes place with plates of glass, or marble, which have been wetted with water.

Exp. 2.—If two globules of quicksilver be placed near each other, they will run together, and become one large drop or ball.

Capillary Attraction.

CAPILLARY ATTRACTION is reckoned a species of cohesion. The suspension of the fluid in the capillary tubes, is owing to the attraction of the ring of glass contiguous to the upper surface of the fluid! and in capillary tubes the heights to which the fluid rises, are inversely as the diameter of the bores.

Exp. 1.—Take a small glass tube open at both ends, dip it in water, and the water will rise in the tube higher than its level in the basin: the water will rise the higher, the smaller the bore of the tube is.

Exp. 2.—Take two pieces of glass, five or six inches square, join any two of their sides, separate the opposite sides with a small piece of wood, so that the surface may form a small opening, and immerse them about an inch deep in a basin of coloured water; then the water will rise between the glasses and form a very beautiful curve.

Exp. 3.—Upon the same principle it is that a piece of sugar, or a sponge, draws up water or any other fluid.

Matter and Motion.

Exp. 1.—Let a boy take in his hand a glass of water, and begin running along, and then suddenly stop; he will find it impossible to keep the water from flying out of the glass. This experiment will afford much amusement, and will impress on the mind the fact, that a body in *motion* has a tendency to continue in that state.

Exp. 2.—If a glass of water be placed on the table, and a boy lay hold of it, and *suddenly* snatch it away, the water will fly out of it on the table; because being *at rest*, it has a tendency to continue so.

Exp. 3.—Place two chairs near to each other, and lay a stick from the back of the one to the other; from the stick suspend by a string a poker or staff, and move it backward or forward against any object. This will afford an idea of a battering ram, and with what ease a vast weight may be moved with great velocity.

MECHANICAL EXPERIMENTS.

Exp. 4.—Lay a twenty-four inch scale, or a long piece of wood divided into twenty-four parts, on a pencil, or any thing else, as a support in the middle; then place a penny at twelve inches distance, and two-pence at six inches distance from the middle, and there will be an equilibrium.

Exp. 5.—Adjust as before, and place three half-pence at eight inches distance, and two at twelve inches distance, and there will be an equilibrium.

Exp. 6.—Vary this experiment in a number of ways, and it will be seen that in proportion as the weight is farther from the centre or fulcrum, the greater weight it will balance on the other side.

Exp. 7.—Attempt to open a large gate by pressing against it near to the hinges; it will be very difficult, or impossible.

Exp. 8.—Attempt the same by pressing against it near the lock, at a great distance from the hinges, it will be quite easy. In the former case it was using a lever of the third, in the latter case it was using a lever of the second kind.

Exp. 9.—Hold out a book at the full length of the arm, and the fatigue will soon be insupportable.

Exp. 10.—Draw in the arm, which is shortening the lever, and the weight may be endured.

Exp. 11.—Let a stick be put through between the bars of a chair, and let two boys lay hold one at each end of the stick; if the chair be suspended from the middle of the stick, they endure an equal weight; but if it be shifted nearer to one of them, he feels a greater weight, and the other feels less; the one having the advantage of a long lever, and the other of a short.

Exp. 12.—Take a piece of string, and suspend from it a ball or weight of any sort, and make it vibrate like a pendulum; then lengthen the string, and cause vibration as before, and it will be seen, that according as the string is lengthened, the vibrations occupy longer time.

Exp. 13.—Take three strings, one 5 inches in length, the other 20, and the third 45; to each of which attach a weight, and make them all vibrate at once; it will be seen, that whilst the longest performs one vibration the shortest performs three; and whilst the second performs one vibration it performs two. The second string being four times as long as long as the first, moves with half the velocity; and the third being nine times as long, moves with one-third the velocity.

EXP. 14.—Take two rods of equal length, and suspend them, and make them vibrate like pendulums: they will perform their vibrations in equal times; but if a weight be fastened towards the middle of one of them, that one will now vibrate much faster; the weight raising the centre of *oscillation,* and having the same effect as diminishing the length.

EXP. 15.—Take two strings of equal length, with a weight attached to the end of each, and make them vibrate; moving the one much farther from the perpendicular than the other, so that it may have to describe a much larger arc; the vibrations will take place in equal spaces of time, the greater velocity compensating for the greater length of the arc

Hydraulics and Hydrostatics.

EXP. 1.—In a cask of water bore holes with a gimlet at different heights, and it will be seen that the water will flow out with different velocities, in proportion to the depth below the surface of the water.

EXP. 2.—Put a cup of water in one scale of a balance, and put an equal weight in the other scale; then lay a piece of wood in the cup, and although it floats on the water, it increases the weight in the scale precisely the same as if it were laid down on the scale.

EXP. 3.—In a vessel of water put a piece of wood, and see how far the water rises, and then weigh it; afterwards take away the wood, and pour in as much water as will rise up to the same mark, and then weigh the vessel; it will be found to be of the same weight as before. This shows that a body floating in water displaces as much as is equal in weight to the body itself, and no more.

EXP 4.—Tie a piece of lead or iron to a piece of wood, so as to make it just sink, and afterwards dissolve salt in the water, and the wood and metal will float. The solution of the salt, increases the specific gravity of the fluid, and makes it bear up a greater weight,

Acoustics.

Exp. 1.—Touch a bell when it is sounding, and the noise ceases.

Exp. 2.—Touch a musical string when it sounds, and the sound will cease, by the vibration being stopped.

Exp. 3.—Hold a musical pitch-fork to the lips when it is made to sound, and a quivering motion will be felt from its vibrations. These experiments show that sound is produced by the quick motions and vibrations of different bodies.

Exp. 4.—Let one person hold his head under water, and another take two stones and strike them together under water, and the sound will be heard by the person whose head is under water.

Exp. 5.—Having stopped both ears with cotton, put your fingers to the teeth of a person who speaks to you, and you will hear his voice.

Exp. 6.—Throw a stone into a pond of water, and observe the circles which proceed from it, and you have a correct representation of the manner in which sound is propagated in circles through the air.

Magnetism.

Exp. 1.—Lay a needle on a piece of glass, or a thin piece of board, and the magnet will still exert its power as before.

Exp. 2.—Suspend a long iron wire horizontally by a thread, then rub it with a magnet, the equilibrium is destroyed, and the one end will dip.

Exp. 3. Put a piece of iron in one scale of a balance, and an equal weight in the other; then bring a magnet under the scale in which the iron is, and it will draw it down.

Exp. 4.—Reverse the experiment, and put the magnet in the scale, and the iron under it, and it will draw it down. These two experiments show that iron attracts the magnet, as much as the magnet attracts the iron.

MECHANICAL EXPERIMENTS.

Exp. 5.—Heat a magnet in the fire, and it will be found to have lost all its magnetical properties.

Exp. 6.—Suspend a magnet nicely poised by a thread, and it will point north and south.

Exp. 7.—Rub several needles with a magnet, and run them through small pieces of cork, then put them to float in a bason of water, and they will all point north and south.

Exp. 8.—Take two magnets, and bring the ends which point to the north near to each other, and they will repel each other.

Exp. 9.—Bring the opposite poles together, and they will attract.

Exp. 10.—Rub a needle with a magnet, and run it through a piece of cork, and put it to float in water; hold a north pole of a magnet near its north pole, and it will keep flying away to avoid it, and it may be chased from one side of a basin to another.

Exp. 11.—Rub four or five needles, and we may lift them as in a string, the north pole of one needle adhering to the south pole of the other.

Exp. 12.—Put a magnet under a piece of paper or glass, and sprinkle iron filings over it, and they will arrange themselves in a very curious manner.

Exp. 13.—Sprinkle iron filings on the magnet itself, and they will arrange in a similar way, a great many being near each end of the magnet, and scarcely any in the middle.

Exp. 14.—Place a needle exactly half way between a north and south pole, and it will not move towards either.

CHEMISTRY.

EXPERIMENT 1.—Dissolve three-quarters of an ounce of Glauber's salts in two ounces of tepid water, pour it while hot into a phial and cork it close. In this state it will not crystallize, even when perfectly cold; but if the cork be now removed, the crystallization will be seen to commence, and proceed with rapidity; affording an instance of the effect of ATMOSPHERIC AIR on CRYSTALLIZATION.

EXP. 2.—Take an ounce of a solution of potass, pour upon it half an ounce of sulphuric acid, lay the mixture aside, and when cold, crystals of sulphate of potass will be formed in the liquor. Here a MILD SALT has been FORMED from a mixture of two *corrosive* substances.

EXP. 3.—Pour a small quantity of strong nitric-acid into a wine glass, add twice its quantity of distilled water or clear rain water, and, when mixed, throw a few pieces of granulated tin into it. A violent effervescence will take place, the lighter particles of the tin will be thrown to the top of the acid, and be seen to play up and down in the liquor for a considerable time, till the whole is dissolved. This is another example of a TRANSPARENT LIQUID holding a METAL IN SOLUTION.

EXP. 4.—Dissolve one ounce of quicksilver *without* heat in three-quarters of an ounce of strong nitric-acid, previously diluted with one ounce and a half of water. Dissolve also the same weight of quicksilver, *by means of heat*, in the same quantity of a similar acid, and then to each of these colourless solutions, add a colourless solution of potass. In one case, the metal will be precipitated in a *black*, in the other, in a *white* powder, affording an example of the difference of colour of metallic oxides, arising from DIFFERENT DEGREES OF OXIDIZEMENT.

EXP. 5.—Fix a small piece of solid phosphorus in a quill, and write with it upon paper. If the paper be now carried into a dark room, the writing will be BEAUTIFULLY LUMINOUS.

EXP. 6.—Take the substance produced in the foregoing experiment, (No. 4), and pour a very little nitric acid upon it. The consequence will be, the solid matter will again be taken up, and the whole exhibit the appearance of one homogeneous fluid. An instance of solid *opake* mass being converted by a chemical agent to a TRANSPARENT LIQUID.

Exp. 7.—Take a transparent saturated solution of sulphate of magnesia (Epsom salts), and pour into it a like solution of caustic potass or soda. The mixture will immediately become almost solid. This instance of the sudden conversion of two fluids to a solid has been called a *Chemical Miracle*.

Exp. 8.—Pour a little phosphureted ether upon a lump of sugar; and drop it into a glass of water, a little warm. The surface of the water will soon become luminous; and if it be moved by blowing gently with the mouth, beautiful and brilliant undulations of its surface will be produced, exhibiting the appearance of a LIQUID COMBUSTION.

Exp. 9.—Put thirty grains of phosphorus into a Florence flask with three or four ounces of water. Place the vessel over a lamp, and give it boiling heat. Balls of fire will soon be seen to issue from the water, after the manner of an artificial firework, attended with the most beautiful coruscations. An experiment to show the extreme INFLAMMABILITY OF PHOSPHORUS.

Exp. 10.—Pour boiling water upon a little red cabbage sliced, and when cold decant the clear infusion. Divide the infusion into three wine glasses. To one add a solution of alum, to the second a little solution of potass, and to the third a few drops of muriatic acid. The liquor in the first glass will assume a purple, the second a bright green, and the third a beautiful crimson. Here is an instance of THREE DIFFERENT COLOURS from the same vegetable infusion, merely by the addition of three *colourless* fluids.

Exp. 11.—Put some prussiate of potass into one glass; into another a little nitrate of bismuth. On mixing these COLOURLESS fluids, a YELLOW will be the product.

Exp. 12.—Prepare a little tincture of litmus. Its colour will be a bright blue, with a tinge of purple. Put a little of it in a phial, and add a few drops of diluted hydrochloric acid; its colour will change to a *vivid red*. Add a little solution of potass, the red will now disappear, and the blue will be restored. By these means the liquor may be changed alternately from red to a blue, and from a blue to a red, at pleasure. An instance of the effects of acids and alkalies in CHANGING VEGETABLE COLOURS.

Exp. 13.—Make an infusion of red roses, violets, or mellow flowers, treat it with solution of potass, and it will become *green*; the addition of diluted muriatic acid will convert it immediately to a *red*. This experiment may be as frequently varied as the last, and furnishes an excellent TEST FOR ACIDS AND ALKALIES.

CHEMICAL EXPERIMENTS. 307

EXP. 14.—Add a drop or two of solution of potass to tincture of turmeric. This will change its original bright *yellow* colour to a dark *brown*, a little colourless diluted acid will restore it. By this tincture we can detect the most minute portion of any ALKALI IN SOLUTION.

EXP. 15.—Pour a little prussiate of potass into a glass containing a *colourless* solution of sulphate of copper, and a REDDISH BROWN will be produced, being a true prussiate of copper.

EXP. 16.—Into a wine glass of water put a few drops of prussiate of potass; a little dilute solution of sulphate of iron into another glass; by pouring these two *colourless* fluids together, a BRIGHT DEEP BLUE COLOUR will be immediately produced, which is the true prussian blue.

EXP. 17.—Prepare a phial with pure water and a little tincture of galls; and another with a weak solution of sulphate of iron; then mix these transparent COLOURLESS fluids together, and they will instantly become BLACK.

EXP. 18.—Pour a little tincture of litmus into a wine glass, and into another some diluted sulphate of indigo; pour these two BLUE fluids together, and the mixture will become perfectly RED.

EXP. 19.—Drop as much sulphate of copper into water as will form a colourless solution, then add a little ammonia, equally COLOURLESS, and an intense BLUE COLOUR will arise from the mixture.

EXP. 20.—If lemon juice be dropped upon any kind of buff colour, the dye will instantly discharge. The application of this acid by means of the block, is another method by which calico printers give the WHITE SPOTS or FIGURES to PIECE GOODS. The crystallized acid is generally used for this purpose.

EXP. 21.—Take a slip of blue litmus paper, dip it into acetic-acid, and it will immediately become Red. This is a test so delicate, that, according to Bergmen, it will detect the presence of sulphuric acid, even if the water contain only one part of acid to thirty-five thousand parts of water. Litmus paper, which has been thus changed by immersion in acids, is, when dried, a good test for the alkalies, for, if it be dipped in a fluid containing the smallest portion of alkali, the red will disappear, and the paper be restored to its ORIGINAL BLUE COLOUR.

EXP. 22.—Write with acetate of cobalt, or with muriate of cobalt, previously purified from iron which it generally contains. When the writing is become dry, these letters will also be invisible. Warm the paper a little, and the writing will be restored to a beautiful BLUE.

Exp. 23.—Draw a landscape with Indian ink, and paint the foliage of the vegetables with muriate of cobalt, the same as that used for making letters green, and some of the flowers with acetate of cobalt, and others with muriate of copper. While this picture is cold it will appear to be merely an outline of a landscape, or winter scene; but when gently warmed, the trees and flowers will be DISPLAYED IN THEIR NATURAL COLOURS, which they will preserve only while they continue warm. This may be often repeated.

Exp. 24.—Write with dilute nitrate of silver, which when dry will be entirely invisible; hold the paper over a vessel containing sulphuret of ammonia, and the writing will appear very distinct. The letters will shine with the METALLIC BRILLIANCY OF SILVER.

Exp. 25.—Write with a weak solution of sulphate of iron, let it dry, and it will be invisible. By dipping a feather in tincture of galls and drawing the wet feather over the letters, the writing will be RESTORED and appear BLACK.

Exp. 26. — Ammonia in solution may very easily be detected by a single drop of muriatic, or acetic acid, which will produce very evident WHITE FUMES.

Exp. 27.—Procure a bladder furnished with a stop-cock, fill it with hydrogen gas, and then adapt a tobacco-pipe to it. By dipping the bowl of the pipe into a lather of soap, and pressing the bladder, soap-bubbles will rise into the atmosphere, as they are formed; and convey a good idea of the principle upon which AIR BALLOONS are inflated.

Exp. 28.—Take a little *red lead*, expose it to an intense heat in a crucible, and pour it out when melted. The result will be metallic glass, and will furnish an example of the VITRIFICATION OF METALS.

Exp. 29.—Prepare two glasses of rain water, and into one of them drop a single drop of sulphuric acid. Pour a little *nitrate of silver* into the other glass, and no change will be perceptible. Pour some of the same solution into the first glass, and a white precipitate of SULPHATE OF SILVER will appear.

Exp. 30.—Prepare two glasses as in the last experiment, and into one of them put a drop or two of *muriatic acid*. Proceed as before, and a precipitate of MURIATE OF SILVER will be produced.

Exp. 31.—Take two glasses as above, and into one of them put a drop of sulphuric acid, and a drop or two of muriatic acid: proceed as before with the *nitrate of silver*, and a MIXED precipitate will be produced, consisting of MURIATE OF SILVER and SULPHATE OF SILVER.

Exp. 32.—Put half an ounce of quicksilver into a wine glass, and pour about an ounce of diluted nitrous acid upon it. The nitrous acid will be decomposed by the metal with astonishing rapidity; the colour of the acid will be quickly changed to a beautiful green, while its surface exhibits a dark crimson; and an effervescence indescribably vivid and pleasing will go on during the whole time the acid operates upon the quicksilver. When a part only of the metal is dissolved, a change of colour will again take place, and the acid by degrees will become paler, till it is as pellucid as pure water. This is one instance of a METALLIC SOLUTION by means of an ACID; in which the opacity of a metallic body is completely overcome, and the whole rendered perfectly transparent.

Exp. 33.—Take the metallic solution formed in the last experiment, add a little more quicksilver to saturate the acid; then place it at some distance, over the flame of a lamp, so as gently to evaporate a part of the water. The new formed salt will soon be seen to begin to shoot into needle-like prismatic crystals, crossing each other in every possible direction; affording an instance of the FORMATION of a METALLIC SALT.

Exp. 34.—If any part of the body be rubbed with liquid phosphorus, or phosphoretted ether, that part, in a dark room, will appear as though it were ON FIRE, without producing any dangerous effect, or sensation of heat.

Exp. 35.—If a little fustic, quercitron bark, or other dye, be boiled in water, the colouring matter will be extracted, and a coloured solution formed. On adding a small quantity of dissolved alum to this decoction, the alumina, or base of the salt, will attract the colouring matter, forming an INSOLUBLE COMPOUND, which in a short time will subside, and may easily be separated.

Exp. 36.—Boil a little cochineal in water, with a grain or two of cream of tartar (supertartrate of potass,) and a dull kind of crimson solution will be formed. By the addition of a few drops of nitro-muriate of tin, the colouring matter will be PRECIPITATED OF A BEAUTIFUL SCARLET. This, and some other instances will give the student a tolerably correct idea of the general process of dyeing woollen cloths.

Exp. 37.—If a piece of calico be immersed in a solution of sulphate of iron, and when dry washed in a weak solution of carbonate of potass, a PERMANENT COLOUR will be produced, viz. the BUFF of the calico printers.

Exp. 38.—Dissolve four drachms of sulphate of iron in one pint of cold water, then add about six drachms of lime in powder, and two drachms of finely pulverised indigo, stirring the mixture occasionally for 12 or 14 hours. If a piece of white calico be immersed in this solution for a few minutes, it will be dyed GREEN; and by exposure to the atmosphere only for a few seconds, this will be converted to a PERMANENT BLUE.

EXP. 39.—Dip a piece of white calico in a cold solution of sulphate of iron, and suffer it to become entirely dry. Then imprint any figures upon it with a strong solution of colourless nitric acid, and allow this also to dry. If a piece be then well washed in pure warm water, and afterwards boiled in a decoction of logwood, the ground will be dyed either of a slate or black colour, according to the strength of the metallic solution, while the printed figures will remain beautifully white. This experiment is designed to show the EFFECT OF ACIDS IN DISCHARGING VEGETABLE COLOURS.

EXP. 40.—Write with a solution of muriate of cobalt, and the writing, while dry, will not be perceptible; but if held towards the fire, it will then gradually become visible; and if the muriate of cobalt be made in the usual way, the letters will appear of an elegant GREEN colour.

EXP. 41.—Put a little oxymuriate of potass and a bit of phosphorus into an ale-glass, pour some cold water upon them cautiously, so as not to displace the salt. Now take a small glass tube, and plunge it into some sulphuric acid: then place the thumb upon the upper orifice, and in this state withdraw the tube, which must be instantly immersed in the glass, so that, on removing the thumb, the acid may be immediately conveyed upon the ingredients. This experiment is an example of a very singular phenomenon, COMBUSTION UNDER WATER.

EXP. 42.—Proceed in all respects as in the last experiment, and add a morsel of phosphuret of lime. Here, besides the former appearance, we shall have COMBUSTION also ON THE SURFACE OF THE WATER.

EXP. 43.—Take the blue solution formed Exp. 19. add a little sulphuric acid, and the colour will *disappear*; pour in a little solution of caustic ammonia, and the BLUE COLOUR WILL BE RESTORED. Thus may the liquor be alternately changed at pleasure.

EXP. 44.—Take water holding carbonate of iron in solution, and add some diluted prussiate of potass; PRUSSIAN BLUE will be formed by the mixture, on adding a few drops of hydrochloric acid.

EXP. 45.—Take some of the same water as that used in the last experiment: boil it, and now add prussiate of potass. In this case NO COLOUR will be produced.

EXP. 46.—Take some water impregnated with carbonic acid, and add to it a little BLUE tincture of litmus. The whole will be changed to a RED.

CHEMICAL EXPERIMENTS. 311

Exp. 47.—Take some of the same carbonated water, and boil it. Then add a little tincture of litmus, and the blue colour will experience NO CHANGE.

Exp. 48.—Introduce a little carbonate of ammonia into a Florence flask, and place that part of the flask which contains the salt on the surface of a bason of boiling water: the heat will soon cause the carbonate of ammonia to rise undecomposed, and attach itself to the upper part of the vessel, affording another example of SIMPLE SUBLIMATION.

Exp. 49.—Fill a glass tumbler half full of lime water; then breathe into it frequently; at the same time stirring it with a piece of glass. The fluid, which before was perfectly transparent, will presently become quite white, and if suffered to remain at rest, REAL CHALK will be deposited.

Exp. 50.—Mix a little acetate of lead with an equal portion of sulphate of zinc, both in fine powder; stir them together with a piece of glass or wood, and no chemical change will be perceptible; but if they be rubbed together in a mortar, the two solids will operate upon each other; an intimate union will take place; and a FLUID WILL BE PRODUCED. If alum or Glauber salt be used instead of sulphate of zinc, the experiment will be equally successful.

Exp. 51.—Write with a solution of nitrate or acetate of lead. When the writing is dry it will be invisible. Then having prepared a glass decanter with a little sulphuret of iron strewed over the bottom of it, pour a little very dilute sulphuric acid upon the sulphuret, so as not to wet the mouth of the decanter, and suspend the writing, by means of the glass stopper, within the decanter. By an attention to the paper the WRITING WILL BECOME VISIBLE by degrees as the gas rises from the bottom of the vessel.

Exp. 52.—Into a large glass jar, inverted upon a flat brick tile, and containing near its top a branch of fresh rosemary, or any other such shrub, moistened with water, introduce a flat thick piece of heated iron, on which place some gum benzoin in gross powder. The benzoic acid, in consequence of the heat, will be separated, and ascend in white fumes, which will at length condense, and form a most beautiful appearance upon the leaves of the vegetable. This will serve as an example of SUBLIMATION.

Exp. 53.—Put a little alcohol in a tea-cup, set it on fire, and invert a large bell glass over it. In a short time an aqueous vapour will be seen to condense upon the inside of the bell, which, by means of a dry sponge, may be collected, and its quantity ascertained. This may be adduced as an example of the formation of WATER BY COMBUSTION.

Exp. 54.—If strong nitrous acid be poured upon a small quantity of a mixture of oxymuriate of potass and phosphorus, FLASHES OF FIRE will be emitted at intervals for a considerable time.

Exp. 55.—Put a little fresh calcined magnesia in a tea-cup upon the hearth, and suddenly pour over it as much concentrated sulphuric acid as will cover the magnesia. In an instant sparks will be thrown out, and the mixture will be COMPLETELY IGNITED.

Exp. 56.—Pour a little water into a phial containing about an ounce of olive oil. Shake the phial, and if the contents be observed we shall find that no union has taken place. But if some solution of caustic potass be added, and the phial be then shaken, an intimate combination of the materials will be formed by the disposing affinity of the alkali, and a PERFECT SOAP PRODUCED.

Exp. 57.—Drop upon a clean plate of copper, a small quantity of solution of nitrate of silver; in a short time a metallic vegetation will be perceptible; branching out in very elegant and pleasing forms, furnishing an example of METALLIC REVIVIFICATION.

Exp. 58.—Make a little charcoal perfectly dry, pulverize it very fine, and put it into a warm tea-cup. If some strong nitric-acid be now poured upon it COMBUSTION and INFLAMMATION will immediately ensue.

Exp. 59.—Dissolve an ounce of acetate of lead in about a quart or more of water, and filter the solution. If this be put into a glass decanter, and a piece of zinc suspended in it by means of a brass wire, a decomposition of the salt will immediately commence, the lead will be set at liberty, and will attach itself to the remaining zinc, forming a METALLIC TREE.

Exp. 60.—Put a table-spoonful of ether into a moistened bladder, and tie the neck of the bladder closely. If hot water be then poured upon it the ETHER WILL EXPAND, and the bladder become inflated.

Exp. 61.—Put a bit of phosphorus into a small phial, then fill it one-third with boiling olive oil, and cork it closely. Whenever the stopper is taken out in the night, LIGHT WILL BE EVOLVED sufficient to show the hour upon a watch.

Exp. 62.—Take a phial with a solution of sulphate of zinc, and another containing a little liquid ammonia, both transparent fluids. By mixing them, a curious phenomenon may be perceived:—the zinc will be immediately precipitated in a white mass, and if then shaken, almost as INSTANTLY RE-DISSOLVED if a little more ammonia be added.

EXP. 63.—Let sulphuric acid be poured into a saucer upon some acetate of potass. Into another saucer put a mixture of about two parts quick-lime, and one of sal-ammoniac both in powder, adding to these a *very small* quantity of boiling water. Both saucers while separate will yield *invisible* gases: but the moment they are brought close together, the operator will be ENVELOPED IN VERY VISIBLE VAPOURS. Muriate of soda, in this experiment, may be substituted for acetate of potass.

EXP. 64.—Take a glass tube with a bulb in the form of a common thermometer; fill it with cold water, and suspend it by a string. If the bulb be frequently and continually moistened with pure sulphuric ether, the water will presently be FROZEN, EVEN IN SUMMER.

EXP. 65.—Dissolve five drachms of muriate of ammonia, and five drachms of nitre, both finely powdered, in two ounces of water. A thermometer immersed in the solution will show that the temperature is reduced below 32 degrees. If a thermometer tube, filled with water, be now suspended within it, the WATER will soon be FROZEN.

EXP. 66.—Pour concentrated nitric acid upon pieces of iron, and a very little action will be seen: but if a few drops of *water* be added, a most violent effervescence will immediately commence, the acid will be decomposed with rapidity, clouds of red nitrous gas will be evolved in abundance, and a perfect SOLUTION OF THE METAL effected.

EXP. 67.—Melt sulphur in a small iron ladle, and carry it into a dark room in the state of fusion. If an ounce or two of copper filings be now thrown in, LIGHT WILL BE EVOLVED.

EXP. 68.—Procure a phial with a glass stopper accurately ground into it; introduce some copper wire, then fill it entirely with liquid ammonia, and stop the phial so as to exclude all atmospheric air. If left in this state, no solution of the copper will be effected. But if the bottle be afterwards left open for some time, and then stopped, the metal will dissolve, and the solution will be colourless. Let the stopper be now taken out, and the fluid will become blue, beginning at the surface, and spreading gradually through the whole. If this blue solution has not been too exposed to the air, and fresh copper filings be put in, again stopping the bottle, the fluid will once more be deprived of its COLOUR, which it will RECOVER ONLY BY THE READMISSION OF AIR. These effects may thus be repeatedly produced.

EXP. 69.—Dissolve some quicksilver in nitric acid, and drop a little of the solution upon a bright piece of copper. If it be then gently rubbed with a bit of cloth, the mercury will precipitate itself upon the copper, which will be completely silvered. This experiment is illustrative of the PRECIPITATION OF ONE METAL BY ANOTHER.

EXP. 70.—If a little nitro-muriate of gold be added to a fresh solution of muriate of tin, both being much diluted with water, the gold will be precipitated of a purple colour, forming that beautiful pigment called PURPLE OF CASSIUS.

EXP. 71.—If a spoonful of good alcohol, and a little boracic acid be stirred together in a tea-cup, and then set on fire, they will produce a very beautiful GREENISH YELLOW FLAME.

EXP. 72.—If alcohol be inflamed in like manner, with a little pure strontian in powder, or any of its salts, the mixture will give a CARMINE FLAME.

EXP. 73.—If barytes be used instead of strontian, we shall have a brilliant YELLOW FLAME.

EXP. 74.—Add a few grains of oxy-muriate of potass to a tea-spoonful or two of alcohol, drop one or two drops of sulphuric acid upon the mixture, and the whole will BURST INTO FLAME, forming a very beautiful appearance.

EXP. 75.—A mixture of oxy-muriate of potass and arsenic, furnishes a detonating compound, which takes fire with the utmost rapidity. The salt and metal, first separately powdered, may be mixed by the gentlest possible triture, or rather by stirring them together on paper with the point of a knife. If two long trains be laid on a table, the one of gunpowder, and the other of this mixture, and they be in contact with each other at one end, so that they may be fired at once; the arsenical mixture burns with the RAPIDITY OF LIGHTNING, while the other burns with comparative SLOWNESS.

EXP. 76.—Into an ale-glass of water put a few pieces of zinc, and a small bit of phosphorus; then drop a little sulphuric acid upon the mixture by means of a glass tube (as described in the experiment of combustion under water), and phosphuretted hydrogen will presently be disengaged, which will INFLAME on rising to the SURFACE OF THE WATER.

J. & W. Rider, Printers, Bartholomew Close, London.

GEOGRAPHY.

33rd Edition, price 3s. 6d., or with Thirty Maps, on Steel, 5s. 6d.,
A SCHOOL GEOGRAPHY,
By JAMES CORNWELL, PH.D.

"Without exception, the best book of its class we have seen. We recommend its immediate adoption by all public and private teachers."—*Atlas.*

Also, price 2s. 6d. plain, 4s. coloured,
A SCHOOL ATLAS,
By JAMES CORNWELL, PH.D

This Atlas consists of thirty beautifully-executed small maps on steel, in which is found every place mentioned in the Author's "School Geography." It also contains a list of several hundred places, with their latitude and longitude. These names are accentuated; and, in cases of difficulty, the pronunciation is also given.

FOR SCHOOLS AND FAMILIES.

14th Thousand, price 5s., crown 8vo., 518 pp., bound in cloth extra
MODERN EUROPE:
A SCHOOL HISTORY.

New Edition, with Three Additional Chapters, comprehending all the Leading Events which have occurred from the Congress of Vienna, in 1815, to the Peace of Villafranca, in 1859. Also copious Questions for Examination. By JOHN LORD, A.M.

"The sketches of character are excellent portraits, and drawn with a free, bold touch. Mr. Lord treats of a portion of history too much neglected in schools."—*Athenæum.*

"The divisions are broad and distinctive, and the style clear."—*Spectator.*

"The book is well and boldly written; great thoughts worthily clothe great facts. The style is a model of historical writing."—*English Journal of Education.*

"Such a book as this has long been wanted for schools."—*Church of England Quarterly Review.*

"Simple, clear, full, and, though concise, minutely accurate, its matter well arranged, its style pure and perspicuous, this volume deserves to be styled a first-rate school book. Everything is supplied necessary for the full comprehension of the history by the reader, and for the best use being made of it by the school teacher, the private governess, or the intelligent parent."—*Patriot.*

SIMPKIN, MARSHALL, & Co., Stationers' Hall Court; HAMILTON, ADAMS, & Co., Paternoster-row; and all Booksellers.

www.ingramcontent.com/pod-product-compliance
Lightning Source LLC
Chambersburg PA
CBHW030738230426
43667CB00007B/758